THE GROWTH OF CHINESE ELECTRONICS FIRMS

The Growth of Chinese Electronics Firms

Globalization and Organizations

Koichiro Kimura

THE GROWTH OF CHINESE ELECTRONICS FIRMS
Copyright © Koichiro Kimura, 2014.

All rights reserved.

First published in 2014 by
PALGRAVE MACMILLAN®
in the United States—a division of St. Martin's Press LLC,
175 Fifth Avenue, New York, NY 10010.

Where this book is distributed in the UK, Europe and the rest of the
World, this is by Palgrave Macmillan, a division of Macmillan Publishers
Limited, registered in England, company number 785998, of Houndmills,
Basingstoke, Hampshire RG21 6XS.

Palgrave Macmillan is the global academic imprint of the above
companies and has companies and representatives throughout the world.

Palgrave® and Macmillan® are registered trademarks in the United
States, the United Kingdom, Europe and other countries.

ISBN: 978–1–137–39140–7

Library of Congress Cataloging-in-Publication Data

Kimura, Koichiro.
  The growth of Chinese electronics firms : globalization and
organizations / Koichiro Kimura.
     pages   cm
  Includes bibliographical references and index.
  ISBN 978–1–137–39140–7 (hardback : alk. paper)
  1. Electronic industries—China.   I. Title.
  HD9696.A3C54455 2014
  338.4′76213810951—dc23   2014019108

A catalogue record of the book is available from the British Library.

Design by Integra Software Services

First edition: October 2014

10 9 8 7 6 5 4 3 2 1

To my wife, Mikiko, and my mother, Yasuko, and in memory of my grandparents, Tosho and Yoshiko

Contents

List of Tables	ix
List of Figures	xi
Preface	xiii
Acknowledgments	xvii
Introduction	1
1 Review and Framework	13
2 Outlook	37
3 Technology Gap	55
4 Diversification Mechanism	77
5 Model	101
6 Challenge for Overseas Expansion	115
Conclusion	131
Notes	139
References	149
Index	169

Tables

2.1	Penetration rates, 1980–2012 (%)	41
2.2	Input coefficients of input–output table, 2010	50
2.3	Trade specialization coefficients, 1995–2013	52
3.1	List of the 38 sectors in the electronics industry	58
3.2	Descriptive statistics for the dataset in 2007	61
3.3	List of variables	61
3.4	TFP levels and growth by ownership	64
3.5	Productivities of indigenous and foreign firms by sector	64
3.6	Results of estimation	69
3.7	Relationship between sector characteristics and inward FDI effects	72
3.8	Sales shares in the top ten firms and inward FDI effects	73
4.1	Market share by major firm, 1999–2008 (%)	84
6.1	Chinese home appliance manufacturers in SA	120
6.2	The top 100 firms in China's home appliance and ICT industry, 2011 (Abbreviated)	122
6.3	SA's import tariff rates, 2011 (%)	123
6.4	OEM businesses of the Chinese investors	126

Figures

2.1 Composition of GDP by industry, 1978–2012 (%)   39

2.2 Market presence of the electrical and electronics industries (a) The ratio of the gross industrial output value of the electrical and electronics industries to the total industrial output value, 1985–2010 (%). (b) The ratio of value addition in the electrical and electronics industries to the total value added of all industries, 1985–2005 (%)   40

2.3 Exports and imports, 2013 (%) (a) Exports (b) Imports   42

2.4 Inward FDI and foreign loans (Flow), 1985–2012 (100 million USD)   43

2.5 Proportion of firms by ownership, 2012 (%)   44

3.1 Relationship between the technology gaps and the ratios of foreign firms' sales to total sales   67

4.1 Market shares of indigenous and foreign firms, 1999–2006 (%)   82

4.2 The number of subscribers, 1990–2008 (Million)   83

4.3 The boundaries of indigenous and foreign firms in the second phase (a) Indigenous firms (b) Foreign firms   85

4.4	The boundaries of indigenous and foreign firms in the third phase (a) Indigenous firms (b) Foreign firms	86
4.5	Product structure	88
4.6	Distribution channels	92
4.7	Product structure (In the case of using the MTK platforms)	98
C.1	Diversification mechanism	133

Preface

The growing presence of developing countries and the understanding of their growth process have become imperative in predicting the effects of the global economy. In contrast, the economic growth in developing countries has been significantly influenced by globalization. An increasing number of developing and developed countries have adopted frameworks that integrate globalization. In the process, many emerging countries have developed growth opportunities. Some of them, including China, have been increasing manufacturing capacity, taking advantage of the abundant labor force. In addition, because of higher demand from industrializing countries, certain developing countries have been scaling up the exports of natural resources. Along with the rapid increase in mining and manufacturing production, their domestic markets have also been expanding. Thus, not only has the growth process of emerging countries been influenced by globalization but these countries have also influenced the global economy in terms of both supply and demand.

This book investigates the influence of globalization on the economic growth in developing countries. Specifically by exemplifying China's electronics industry, it examines how indigenous firms in developing countries have been growing in congruence with globalization. Indigenous Chinese manufacturers have realized rapid growth in the face of globalization. In the rapidly growing manufacturing sector, China's electronics industry has become a leading one. However, although indigenous firms have acquired technologies during the economic liberalization, they have been still facing a technology gap from foreign firms. Nevertheless, their growth has not been impeded by the entry of competitive foreign firms in the Chinese market. Indigenous firms have attained growth by outsourcing technologically advanced production processes to specialized firms and by focusing on marketing processes in which they have some advantages as local firms. Consequently, they have been growing by choosing a unique strategy in comparison with their rivals.

The growth process of indigenous firms is, therefore, complicated. Indigenous firms need to satisfy some conditions to absorb technology from developed countries, because generally technology is not automatically spilled out. Moreover, they have to offset the technology gap in such a way that technologies do not diffuse entirely. To overcome the disadvantage, they need to find measures other than technology acquisition. This is because investment in highly advanced technologies may prove futile if indigenous firms have not assimilated existing technologies. Consequently, indigenous firms always need to strike an "optimal balance" between technology acquisition and the measures to reduce the technology gap. You might project the scenario that Chinese firms have been enjoying absorbing technology and have realized rapid growth. The reality, however, is not quite so. When I visited Chinese firms, I almost always heard that they were facing a dearth of core technologies. Therefore, Chinese firms are required to find measures to compensate for the technology gap at the same time. This study investigates the behavior of Chinese firms to elucidate certain crucial characteristics of indigenous firms in developing countries.

To achieve further growth in the global market, Chinese firms must accumulate more advanced technologies to improve their competitiveness on basis of their experiences in the home market. This is a challenge to Chinese firms, but not only to them. Indigenous firms in other developing countries also face more or less the same challenge. This merits the need to examine the growth processes of indigenous firms in various developing countries to predict the global economies. This book is written and compiled with the hope of providing its readers with the requisite knowledge to comprehend the growth of indigenous firms and the future of the global economy.

This study examines the growth of indigenous Chinese electronics firms in the era of globalization. Some conclusions have been presented in my PhD dissertation, and certain chapters of this book were published in journals referred hereinafter. Since the conclusions of the articles were presented in piecemeal, this book is an organized presentation of the study and its conclusions.

Acknowledgments

I would not be able to finish this book without the warm support of many people. I would like to express my gratitude to Professor Yukiko Fukagawa (Waseda University), the late Professor Kazuharu Kiyono (Waseda University), Professor Hideki Konishi (Waseda University), and Professor Zhijia Yuan (Rissho University) for their advices and comments on my PhD dissertation. Also, I am hugely grateful to Emeritus Professor Reiitsu Kojima (Daito Bunka University) and the late Emeritus Professor Sadao Suwa (Waseda University) for their advices and comments on my BA and MA theses, respectively.

This study was supported by the Institute of Developing Economies, Japan External Trade Organization (IDE-JETRO), and Grant-in-Aid for Scientific Research in Japan. I also express my gratitude to the following individuals for their comments and giving me the opportunity to join their research projects: the late Mr. Kenichi Imai (IDE), Dr. Momoko Kawakami (IDE), Professor Tomoo Marukawa (University of Tokyo), Professor Yasunori Ishii (Waseda University), Professor Ichiroh Daitoh (Keio University), Associate Professor Nobuhiro Horii (Kyusyu University), Associate Professor Moriki Ohara (Ryukoku University), Professor Mariko Watanabe (Gakushuin University), Dr. Kumiko Makino (IDE), and Dr. Chizuko Sato (IDE). In addition, I would not be able to finish my articles without my field research with Professor Yuan, the late Mr. Imai, Dr. Kawakami, Professor Marukawa, Mr. Hideto Akiba

(Ministry of Economy, Trade and Industry of Japan), Professor Masanori Yasumoto (Yokohama National University), Associate Professor Hirohumi Tatsumoto (University of Tsukuba), Mr. Jing-Ming Shiu (University of Tokyo), and Dr. Takahiro Fukunishi (IDE). I humbly thank them.

Part of this book is based on research conducted at Institute of Industrial Economics, Chinese Academy of Social Sciences (IIE-CASS), as a visiting research fellow during the two years from June 2005 to 2007. I tender cordial thanks to former President Zheng Lü, Dr. Ying Zhao, Dr. Xiaoxia Xie, Dr. Jun He, Ms. Yi Ding, and research fellows at the institute for their advice.

I would like to thank all the research project members; the discussants and questioners present at academic meetings held in Japan, China, and the United States, where I presented my papers; the editors and anonymous referees of the journals undermentioned, and co-workers at IDE-JETRO for their comments and supports. I also thank all the interviewees at firms and related organizations for their cooperation and responding to my questions despite their busy schedule.

As noted in Preface, this book includes articles published in journals. I wish to thank the publishers for giving me the permission to use the copyrighted materials. The following articles were published by Wiley-Blackwell, Taylor & Francis Group, and the Graduate School of Economics, Waseda University, respectively.

Kimura, Koichiro (2012) "Does Foreign Direct Investment Affect the Growth of Local Firms? The Case of China's Electrical and Electronics Industry," *China & World Economy* 20(2): 98–120.

Kimura, Koichiro (2011) "Is There Hope for Firms Facing the Technology Gap? A Case of China's Mobile Industry," *Journal of Contemporary China* 20(72): 833–847.

Kimura, Koichiro (2011) "Gijutsu Kakusa to Kigyo no Kyokai [The Technology Gap and the Boundaries of the Firm]," *Waseda Economic Studies* 70: 1–16 (in Japanese).

The first, second, and third articles have been included in chapters 3, 4, and 5, respectively, after certain revisions.

I am thankful for the support of the editors, Ms. Leighton Lustig, Ms. Leila Campoli and Ms. Sarah Lawrence, for giving me the opportunity to publish this book.

Finally, I am extremely grateful to my family; my wife who has always supports me, and my mother and grandparents who created opportunities by providing me with my education.

INTRODUCTION

I.1 RESEARCH OBJECTIVE

Globalization has made an inescapable impact on almost every country, even those temporarily adopting a closed-door policy. At the same time, every country has influencing globalization, albeit at varying magnitudes. Although developing countries are generally more affected, their own influence on the global economy is growing rapidly. This apparent interaction necessitates that we first understand how a developing country grows in a situation of globalization. If we do not understand this, we would not be able to predict how the country would affect the global economy. This book investigates the influence of globalization on developing countries at the firm level; specifically, the influence of foreign firms from developed countries on the growth of indigenous firms in developing countries, by exemplifying China's electronics industry.

There are two key terms for this book, indigenous and foreign firms, which we define here. Indigenous and foreign firms are firms in developing and developed countries, respectively.[1] In this study, indigenous firms are organizations established in their respective countries and are owned by the majority of local capital.[2] And foreign firms are organizations in developed countries, which sell their products in developing countries, regardless of exports or investment being the mode of overseas expansion. Based on this definition, we examine the influence of foreign firms on the growth of indigenous firms.

The influence of globalization on developing countries and their firms is scrutinized so often because this argument is used to explain the possibility of growth in developing countries and their firms.[3] Because globalization increasingly links developing and developed countries via factors such as international trade and foreign direct investment (FDI), developing countries and their firms cannot avoid the influence of developed countries and their firms in terms of both positive and negative effects of globalization. A significant difference between developing and developed countries, or indigenous and foreign firms, is their existing levels of technology. Technology is defined as a set of knowledge required for production (Mansfield, 1968). Indigenous firms resort technology acquisition to catch up with foreign firms; whereas foreign firms already have such technologies because their operations initiated much earlier than those of indigenous firms; nevertheless, foreign firms tend to seek new technologies to further their growth. To this effect, foreign firms can become technology providers for indigenous firms, or in some cases, tough competitors. However, predicating the effect of a priori is not possible since there is a possibility of either. Therefore, the question of whether developed countries and their firms have a positive or negative effect on developing countries and their firms has been a long-standing issue that has attracted much attention.

Numerous studies have evaluated the effect of globalization on developing countries and their firms. In a related research field, technology diffusion has been identified as a significant determining factor for the positive or negative effect of globalization.[4] In this study, technology diffusion is defined as the spread of technologies in developed countries to developing countries through various channels, such as technology transfers, technology introductions, and technology imports. Put differently, technology diffusion is the equalization of the technological levels in developing and developed countries, or indigenous and foreign firms, by absorbing technologies, irrespective of the said channels. Studies on the effect of globalization have recognized that the growth in developing countries and their firms depends on technology diffusion. Specifically, if technologies diffuse from developed countries and their firms to developing countries and their firms through international economic activities, then developed countries and their firms can promote the growth in developing countries and their firms through the positive effect, that is, the technology spillover effect. However, if technologies do not diffuse for some reason, then competitive and productive foreign firms would impede the growth of indigenous firms and would dominate fast-growing markets in developing countries, causing a negative effect, that is, the market-stealing effect (Aitken and Harrison, 1999). In this way, growth in developing countries and of their firms depends on whether or not they can get level at the technological level with developed countries and their firms.

Some examples of success are the East Asian economies, such as South Korea, Taiwan, Hong Kong, Singapore, Thailand, Malaysia, and so on, which have been generally receiving a positive effect of globalization. They have been developing by opening up their economies to the world and by absorbing technologies from developed countries. Some of them initially adopted the strategy of import-substitution industrialization (ISI) in the same way as a number of countries across the world adopted this approach for industrialization in the post–World War II era. But when domestic markets got saturated, these East Asian countries quickly shifted their strategy from ISI to export-oriented industrialization (EOI). As a result, many East Asian economies have been successful in boosting their growth. Openness provides avenues to diffuse technologies, which are an important factor for economic growth, and to realize convergence of the economic levels.

Following the success stories of East Asian economies, China launched its economic reform and open-door policy in 1978 and embarked on a path of rapid growth. Although China had almost closed its economy to the world in its planned economy era, international trade and inward FDI flow have been increasing ever since economic liberalization was started. Therefore, it can be said that the Chinese economy has been growing through economic internationalization. In the rapidly growing economy, the manufacturing sector has been developing significantly. Moreover, a number of indigenous Chinese firms have been rapidly growing and have become tough competitors for foreign firms in the Chinese market. Although there are many countries mainly led by foreign firms while the economies are growing, indigenous Chinese firms have become market leaders in many manufacturing industries within China. Our evaluation is that China possibly has been, at least on the whole, enjoying the positive effect of globalization thus far.

In particular, the electronics industry has become a leading and representative one in "the world's factory." A number of Chinese electronics firms have been growing rapidly. In this book we mainly include consumer electronics products and home appliances in the electronics industry. Of course, white goods, such as refrigerators, washing machines, and such home appliances basically belong to the electrical industry, not to the electronics industry. However, this book hereinafter collectively refers to them as the electronics industry, because home appliances manufacturers are substantially overlapping with electronics manufacturers in the market. Nevertheless, when we have data separately classified in the electrical and electronics industries, we show figures for each based on the industrial classifications.

While Chinese electronics firms have been growing significantly, there is a profound difference in the organizational form between Chinese and foreign firms even within the same industry. In general, production includes many goods and services that are part of a long value chain of the product. Therefore, the decision of whether or not a firm integrates a stage involved in making goods and services, that is, the so-called make-or-buy decision, varies by firm. To understand the contrast in organizational form, let us divide the value chains of electronics products into the three stages: development, manufacturing, and sales. First, product development on the upstream of the value chain is a stage to design and develop new models, and also other related activities. Next, product manufacturing on the midstream is a stage to manufacture and assemble products after procuring their components, and also other related activities. Finally, the product sales on the downstream are a stage to sell products and conduct the after-the-sale services, and also other related activities.

Unlike foreign firms, Chinese firms have tended not to integrate a great part of the development stage and part of manufacturing stage, because they do not have core technologies to design and develop new models and to manufacture key components though technologies have been matured to some extent. They also, of course, have been making efforts to absorb technologies from foreign firms, which is one of the benefits of globalization, by accumulating through production, and by developing through investing in long-term research and development (R&D) activities. Among them, technology diffusion from foreign firms into indigenous firms is perhaps most significant for any indigenous firm at the stage shortly after its entry into the industry. However, it may not be possible that in this way they acquire all the technologies they need. Even if indigenous firms try to acquire all of them through absorption, the firm would not be efficient enough to acquire more advanced technologies before assimilating the technologies it already possesses. There are limitations to technology acquisition in a short span of time. Therefore, Chinese firms have been experiencing a technology gap, and are still facing it, while competing with competitive foreign firms. This may be termed the technology disadvantage and advantage for indigenous and foreign firms, respectively.

To offset the perpetuating technology gap, indigenous firms have tended to use the following two makeup measures. The first measure is buying technologically advanced goods and services for the development and manufacturing stages from outside specialized firms. The demarcation of each of the development and manufacturing stages from other stages can be effected by developing a vertical specialization of industrial structures using digitalization of products and modularization of product structures. In this book, the vertical specialization describes a situation wherein some goods and services concerning a product are made by outside specialized firms.[5] It is well known that firms have been able to produce products with basic functions as long as they can buy technologically advanced goods and services from outside firms. Therefore, firms are not required to develop new models or to manufacture key components from the beginning by themselves. The development of vertical specialization has made it easier for latecomers seize the opportunity to enter and grow.

Another measure consists mainly of internalizing part of the sales stage in comparison with foreign firms. Because indigenous firms are familiar with the domestic market, they enjoy some advantages in managing local sales staff, to establish an intricate sales-and-service network and expand it, to find reliable local distribution partners, to understand demand trends and business practices in China, and so on. Therefore, Chinese firms have established nationwide sales and after-the-sale service networks, and have planned such new models as to meet the demands of the Chinese market. There is a potential for indigenous firms to sell products more effectively than foreign firms, but the potential for them to develop products more effectively than foreign firms is not there. This can be the home advantage in marketing and the away disadvantage, or overseas disadvantage, in product development for indigenous and foreign firms, respectively. As a result, since the advantage and disadvantage of Chinese firms show a contrast with those of foreign firms, the make-or-buy decisions of Chinese firms also show a contrast with those of foreign firms.

Thus, globalization neither necessarily makes the technological levels of indigenous and foreign firms uniform through technology diffusion nor does it necessarily inhibit the growth of indigenous firms with the entry of competitive foreign firms into developing countries' markets. On the contrary, globalization rather has a force to diversify the organizational structure of indigenous firms by stimulating them to build on what they are good at, than by trying to overcome what they are not good at. Chinese electronics firms have sought to find an optimal balance among the three factors, which we can call as follows: the technology gap, external technology, and internal knowledge. Since technologies are also knowledge, as Mansfield (1968) stated, therefore it is, as it were, global knowledge, which works transnationally by forming the core functions of products. In contrast, we cannot define the home advantage in marketing based on narrowly defined technologies; we can rather define it in terms of local knowledge connaturally owned by indigenous firms. Of course, part of global knowledge can work only in specific local areas for customizing products to adapt to relevant local areas. On the other hand, part of local knowledge can work not only in specific local areas but also in the global market because of the general versatility of the knowledge. However, the difference between the natures of global and local knowledge generally determines the difference between indigenous and foreign firms in terms of their advantages and disadvantages. Consequently, when we call technologies knowledge, it turns out that indigenous firms face problems to reconcile among any possible lack of global knowledge, the accessible global knowledge, and the local knowledge.

Previous studies have not systematically demonstrated the diversification mechanism of organizations, although each of the three factors has been studied independently. First, the relationship between indigenous and foreign firms has not been explicitly analyzed in terms of organizational form, even though the existence of foreign firms has a significant influence on the organizational diversification of indigenous firms in the era of globalization. The dissimilar advantages and disadvantages of indigenous and foreign firms, the contrast can serve to diversify the organizational form of firms through competition between the two sides. Second, the rational selection for a balance between integration and disintegration of business stages has not been analyzed accurately, even though indigenous firms tend to cause the development stage disintegrate and tend to integrate the sales stage. However, indigenous firms do not buy all of development services and, similarly, do not internalize all of marketing services. Therefore, there must be an optimal balance between complete integration and disintegration. We will investigate the influence of foreign firms on the growth of indigenous firms, so that we can elucidate the diversification mechanism.

1.2 STRUCTURE

This book aims to develop the idea of a diversification mechanism through a case of China's electronics industry. It consists of the following eight chapters, including the Introduction and Conclusion. The chapters can be categorized into four parts, though it is not explicitly denoted. The first part includes the Introduction and Chapter 1. In Chapter 1, the question of this book overviewed in the previous section will be discussed again in more detail. In addition, an analytical framework and methods used in this book will also be explained there.

The second part includes Chapters 2 and 3. This part will enunciate the preconditions to develop the diversification mechanism by studying China's electronics industry. Chapter 2 justifies why this study focuses on the electronics industry by providing three reasons and provides an overview of the industry's development. The first reason is that the electronics industry is a leading and representative one in the fast-growing manufacturing sector in China. Since Chinese electronics firms have attained rapid growth despite facing a technology gap, the case can be a lesson for indigenous firms in other developing countries for globalization. The second reason is that both indigenous and foreign firms are growing in China's electronics industry. Therefore, we can analyze the competition between indigenous and foreign firms in the Chinese market in order to investigate the difference in organization between the two sides. The final reason is that the vertical specialization of industrial structures is well developed in the industry. In other words, because of vertical specialization, firms can choose to internalize a stage to produce goods and services in-house or to buy it from an outside firm, thus developing various patterns of organizational forms in a value chain.

Chapter 3 empirically verifies the lack of technology diffusion from foreign to Chinese firms in certain sectors, in which the technology gap is relatively large in China's electronics industry. To explain it, the chapter shows that inward FDI does not necessarily diffuse technologies and does not necessarily have the positive influence of globalization on the growth of indigenous firms. Specifically, we investigate whether inward FDI has the positive or negative influence on the value-addition and total factor productivity (TFP) of indigenous firms in 38 sectors of the electronics industry. As a variable of inward FDI, a ratio of fixed assets of foreign firms to total fixed assets in each sector is used. The results show that sectors with less experience and lower technological levels of indigenous firms, in comparison with those of foreign firms, are likely to experience the negative effect of globalization. Consequently, although indigenous firms in these sectors must have greater growth potential, they cannot absorb enough technologies from foreign firms due to the significant gap and the lack of experience of operations. Therefore, even though the industry has been developing as a whole, the effect of inward FDI depends on the characteristics of the sectors. However this study does not insist that technologies did not diffuse; rather it highlights certain limitations to technology diffusion and explores the negative influence.

The third part includes Chapters 4 and 5. This part will explain how Chinese electronics firms have attained growth by striking a balance among the following three factors: the technology gap, external technology, and internal knowledge. In Chapter 4, we will concretely investigate the diversification mechanism through a case of China's mobile phone handset industry. Although the product structure was partially modularized, Chinese handset firms faced a technology gap that could not be closed because indigenous firms had not accumulated enough experiences to develop and design products efficiently. However, they realized rapid growth by aligning with outside firms that provided development and design services and key components, and by using the home advantage of the ability to establish nationwide sales networks and knowing the kind of new models that the Chinese market demands. Using external technology had the rationalities that outside firms had already accumulated enough experiences, and that product differentiation was not so much required in the Chinese handset market at the time. Using internal knowledge also had the rationality of finding business opportunities appropriate for the fast-growing market. They realized growth by selling inexpensive handsets in local and rural markets, which foreign firms did not focus on.

Chapter 5 presents the development of a model to illustrate the essence of the previous chapter and to generalize the case study. Specifically, we incorporate the concept of the technology gap into the models on the make-or-buy decision. Our results are as follows. Even though it is optimal to make technologically advanced services in-house in terms of the theory of the make-or-buy decision, indigenous firms may not be able to realize this ideal situation if advanced technologies are required. As a result, indigenous firms may choose to buy the services from outside firms, because it is more rational than making them in-house. However, if there are technological changes in the product structure, and/or if indigenous firms in developing countries can effectively exert the home advantage, the chances of the firms to enter and grow may increase. By developing the mechanism, we extract entry conditions to balance the three factors. Moreover, when comparing results for the make-or-buy decision, the results for the model with a technology differ from those for the model without it.

The fourth part includes Chapter 6 and the Conclusion. In this part, we will apply and examine the conclusions on the diversification mechanism that merits further analysis on the future growth of Chinese firms. Chapter 6 will discuss a challenge of Chinese electronics firms through a case of outward FDI in South Africa (SA). The outward FDI from China is increasing in recent years, with the promotion of "Going Global (*Zouchuqu*)" strategy by the Chinese government. However, since Chinese firms have been growing in the home market by utilizing internal knowledge and have been isolated from foreign markets, expecting them smoothly grow in the foreign markets is not possible. It is because there are a lot of competitive rivals from developed and host countries. Therefore, Chinese firms have been expanding their businesses through transactions with indigenous SA firms that have markets, in order to compensate for the lack of awareness about Chinese firms' brands in the foreign market. We show that the diversification mechanism can shed light on the Chinese firms' strengths in the home market and on the weakness or a challenge in away markets.

The Conclusion examines the concept of the diversification mechanism. The organizational form can be diversified by striking a balance between the technology gap, external technology, and internal knowledge. Depending on these three factors, this study shows three possible patterns for the growth process of indigenous firms in terms of organization. Each pattern is as follows: homogenization with foreign firms, diversification in comparison with foreign firms, and zero organizations. The first emerges by (almost) complete technology diffusion. And the last case means that indigenous firms cannot enter and grow despite a significant technology gap that cannot be compensated with external knowledge and internal knowledge. The diversification mechanism can cover not only the diversification case but also the others, that is, homogenization and zero organizations. In addition, while the indigenous firms infer that their diversification mechanism has developed, we can expand the conclusions to the interaction between indigenous and foreign firms. If foreign firms also learn the strengths of indigenous firms, the standard of knowledge needed to compete with rivals in the industry can rise. Although we have only focused in this book on the growth of indigenous firms, the diversification mechanism can lead to a development mechanism for the whole of firms under competition or for the whole of an industry if indigenous firms can be competitive and foreign firms also intend to learn the strengths of indigenous firms.

CHAPTER 1

REVIEW AND FRAMEWORK

1.1 INTRODUCTION

Numerous studies on the influence of globalization on developing countries and their firms have already been conducted in international economics and related research fields, and, as we will see in this chapter, a variety of viewpoints have been expressed on this influence. There, however, still remain questions to be answered. One of these questions concerns the growth of indigenous firms in developing countries when there is incomplete technology diffusion.[1] Although technology diffusion from developed countries and their firms to developing countries and their firms has been the focus of several studies, to the best of the author's knowledge partial diffusion of technology has not been investigated explicitly. This study conducts an in-depth analysis of incomplete technology diffusion because it is more commonly experienced in reality.

Studies on the modern Chinese economy have analyzed the growth of Chinese firms with such a technology gap. These studies have shown that Chinese firms tend to purchase technologically advanced goods and services from outside firms and focus mainly on their marketing. However, existing research has not explicitly considered the influence of foreign firms on the make-or-buy decision of Chinese firms.

Consequently, little is known about the rationale underlying Chinese firms' decision in choosing various organizational structures to tackle competition from foreign firms. Thus, the present research question on the diversification of organizations remains unsolved.

To answer this question, we study the diversification mechanism by integrating the existence of the technology gap into the make-or-buy decision; that is, we study the boundaries of indigenous firms in developing countries. The boundaries of the firm describe the integrated stages which goods and services are produced in-house (Besanko et al., 2009). Firms decide on these boundaries in such a way as to maximize the effect of investment in the integrated stages, as we discuss later. Focusing on the organizational level of the firm enables us to analyze the diversification within the firm.

In this chapter, we review the literature in this area of study and then put out our research question. We then present an analytical framework and the methods to solve the question.

1.2 REVIEW

The ensuing subsections review exsting studies on the influence of developed countries and their firms on the growth of developing countries and their firms, and the influence of those on the growth of Chinese firms.

1.2.1 THE INFLUENCE OF DEVELOPED COUNTRIES ON DEVELOPING COUNTRIES

1.2.1.1 Globalization and Growth: Model Analysis
The influence of globalization on the economic growth in developing countries has long been a subject of study.[2] It has been studied from various perspectives since the development of endogenous growth theory in the 1980s (Bardhan and Udry, 1999; Jensen and Wong, 1997; Long and Wong, 1997; Todo, 2008), though prior to that it was analyzed mainly using comparative statics (Meier, 1963). Endogenous

growth theory explicitly explains technological change as a factor in long-term growth (Acemoglu, 2009; Aghion and Howitt, 2009; Barro and Sala-i-Martin, 2004; Grossman and Helpman, 1991), whereas this phenomenon is not explained in neoclassical growth theory (Solow, 1956). To include technological change, the endogenous growth theory incorporates technological improvements, such as learning-by-doing (LBD), which increase the productivity of research and development (R&D) (Romer, 1986); human capital investment through education (Lucas, 1988); an increase in the variety of new goods (Romer, 1990); and others.

Studies on the relationship between globalization and growth have been advanced by extending endogenous growth theory to open-economy models. On one hand, Grossman and Helpman (1991) showed that international trade between developing and developed countries results in economic growth in developing countries through technology diffusion from developed countries to developing ones. Conversely, openness does not necessarily result in economic growth when technologies do not diffuse for some reason. Young (1991) analyzed the negative effect of globalization. According to his model, LBD leads to the technology gap—in other words, the productivity gap—between developing and developed countries. Developed countries can specialize in manufacturing high-tech products that have room for productivity improvement. On the contrary, developing countries have no other alternative but to manufacture low-tech products that do not have room for productivity improvement. Therefore, globalization prevents the development of high-tech industries in developing countries.

This reveals that technology diffusion determines the impact of globalization on industrialization. The results of trade between countries with varying technological levels (i.e., between a developing country and a developed one) provides results contrary to those produced from international trade between countries with similar technological levels (i.e., between developed countries.)[3] Rivera-Batiz and

Romer (1991) showed that trade between developed countries can increase the knowledge base available for R&D and can enhance the economic growth of both countries involved. However, their model also suggests that when technologies do not diffuse, trade between a technologically advanced country and one which does not have enough technological capabilities can impede the economic growth of the latter. Thus, varying technological levels create different effects on developing and developed countries.

1.2.1.2 Technology Diffusion
Several studies have been conducted on technology diffusion as an essential growth factor and its positive effect. Mansfield et al. (1982) have argued for the positive effects of technology diffusion. According to the authors, 26 technological items developed in the United States (US) diffused to indigenous firms abroad through US overseas subsidiaries for an average of four years from 1960 to 1978. The diffusion contributed toward decreasing production costs incurred by indigenous firms in the host countries.

Such an effect—that is, the increase in productivity in developing countries through technology diffusion—is also known as the advantage of backwardness. Drawing on the study of the experiences of European countries in the nineteenth century, Gerschenkron (1962) showed that industrialization can occur at an accelerated pace if developing countries succeed in satisfying the preconditions necessary for economic advancement. The preconditions include measures such as intensified investment using banking systems and aggressive interventions by governments to promote industrialization. When developing countries satisfy the preconditions, they can effectively absorb new technologies and attain rapid economic growth. Therefore, Gerschenkron does not describe an automatic diffusion of technology. Moreover, he showed that the development process in developing countries can be different from that in developed countries. This idea is closely linked to our discussion.

In addition to the European cases, in the late twentieth century East Asian economies have also enjoyed the advantage of backwardness. The presence of a variety of preconditions, including the role of governments, has been emphasized in studies concerning the development of East Asia. Amsden (1989) demonstrated that South Korea (Korea) adopted aggressive protectionist policies to their domestic industries, especially large conglomerates (*chaebol*), and provided opportunities for indigenous firms to absorb technologies from developed countries. In addition to these factors, Fukagawa (1997) focused on the emerging economic agents of labor (labor unions) and financial institutions (especially banks) and showed that Korea's rapid development was through the absorption of many technologies, particularly from the US and Japan.[4] Kim (1997) and Kim and Nelson (2000) argued that Korean firms rapidly advanced from the stage of imitation to that of innovation through technology diffusion. In a study on Asian newly industrializing economies (NIEs), Watanabe (1979) found that developing countries can steadily grow when they have social ability, including the existence of abundant skilled labor, entrepreneurial abilities, and governmental administrative abilities. Suehiro (2000) analyzed the case of Thailand with regard to social ability and found that it led to advancement of industrialization in both the government and private sectors. Social ability also has a strong relationship with governments' involvement in their economies. Similarly, other East Asian emerging economies also satisfied the preconditions and seized the advantage of backwardness.

Technologies can diffuse through not only international trade but also inward foreign direct investment (FDI) (Keller, 2004).[5] On one hand, imports provide the opportunity for indigenous firms to have exposure to the innovations of, and new products developed by, foreign firms, through which they can expect to gain knowledge of new technologies. On the other hand, exports provide the opportunity for indigenous firms to compete with other aggressive firms in the global market. In the process, indigenous firms can expect to increase their productivity through exports or the so-called learning-by-exporting. Inward FDI also creates opportunities for technology diffusion through the operations of multinational enterprises (MNEs) in host countries. For example, employees who switch their place of employment from MNEs to indigenous firms can spread learning in their new workplaces. Various studies have analyzed international trade and FDI as diffusion channels of technologies.

Transactions and affiliations with foreign firms in developed countries can also become effective channels of technology diffusion. Hobday (1995) revealed that indigenous firms in Asian NIEs—that is, Korea, Taiwan, Hong Kong, and Singapore—acquired technologies through these channels. Moreover, Kawakami and Sturgeon (2011) found that East Asian firms in various industries joined global value chains (GVCs) and received opportunities to learn technologies from leading foreign firms.

In addition, technological obsolescence also can lead to technology diffusion. Based on the lifecycle of products (introduction, growth, maturity, and decline of products), Vernon (1966) hypothesize the product lifecycle (PLC) and that production bases shift to developing countries when products produced in developed countries become obsolete. Krugman (1979) developed a formal model of the hypothesis, according to which new products produced in developed countries can be imitated by developing countries. It showed that patterns of trade in new and existing products between developed and developing countries depend on the speed of new product development by developed countries and the speed of imitation of those new products by developing countries.

The PLC hypothesis has also been employed to analyze the impact of globalization on developing countries. Grossman and Helpman (1991) developed an endogenous growth model to explain the economic growth in developing countries by exemplifying a cat-and-mouse game to elucidate technological progress in developed countries and imitation in developing countries. Antràs (2005) incorporated the PLC hypothesis into the boundaries of the firm. He showed that whether or not manufacturers in developed countries decide to outsource to developing countries depends on the extent of technological obsolescence.

Technologies can diffuse through a variety of channels. However, sometimes they do not diffuse due to various reasons. The first reason is the characteristics of technologies. If technologies are developed through implicit knowledge and know-how, it is difficult to access them by definition. For example, LBD also demonstrates such characteristics. Since technologies that result from LBD can be acquired only through production and business experience, the benefit of LBD is restricted to the learners themselves. In addition, if technologies are protected by patents, then they are virtually restricted from diffusion. Although access to technologies can be easier due to the disclosure of technologies through the patent system, the use of those technologies is regulated and entails high costs in some cases.

The second reason for the inhibition of technology diffusion is closely connected to the ability to absorb new technologies. Backwardness is not the only condition for enjoying technology diffusion; developing countries are also required to accumulate the ability of technology absorption. Keller (1996) indicated that a certain level of R&D investment by indigenous firms is required to increase the ability to absorb technology. Therefore technology diffusion is a conditional phenomenon and globalization cannot be defined as wholly positive or negative.

The third reason corresponds to the degree of the technology gap. The technology gap is an advantage of backwardness for developing countries; however, it can become a disadvantage if the gap is too large to close. Aghion and Griffith (2005) found when technologies possessed by foreign firms are too advanced for indigenous firms to learn, the indigenous firms may not be able to absorb those technologies.

The final reason that inhibits diffusion is related to historical backgrounds. Parente and Prescot (2000), based on their study of India's textile industry before World War II, stated that labor resistance against the introduction of new technologies often blocks technological progress and decreases income levels. Although many developing countries and their firms have been absorbing technologies, we cannot ignore these preconditions for realizing the advantage of backwardness.

1.2.1.3 Globalization and Growth: Empirical Analysis

Since theoretical considerations have yielded opposing results, empirical analyses have been conducted in the attempt to verify whether globalization has a positive or negative effect. However, it is difficult to draw a conclusion because of methodological limitations.

Although many studies have reported the positive effects of international trade on the economic growth in developing countries, methodological problems also have been pointed out (Harrison and Revenga, 1995; Harrison and Rodríguez-Clare, 2010; Rivera-Batiz and Oliva, 2003). Economic growth rates are generally regressed on a variable of openness; however, researchers are immediately faced with the question of what would be the most appropriate indicator of openness. Some studies used the difference between domestic and international prices; however, the proxy may just reflect macroeconomic situations, such as foreign exchange rates. To deal with the effect of trade policies explicitly, tariffs have been often used; however, there still remains the question of whether tariff revenues or tariff rates should be used. In addition to the indicators relating to trade policies, the ratio of trade amount to GDP has also been used, but its usage may create an endogeneity problem between income of a dependent variable and trade amount of an independent variable.

In addition to studies on the relationship between trade and growth, empirical studies on the relationship between inward FDI and growth have also been increasing; however, the results are ambiguous. Using data on Mexico, Kokko (1994) showed that inward FDI can have a positive effect of technology diffusion. In contrast, Aitken and Harrison (1999) showed that foreign firms have possibilities to dominate markets, as exemplified by the negative market-stealing effect on the productivity of indigenous firms in Venezuela. Also for the Czech Republic, Djankov and Hoekman (2001) showed that inward FDI can have a negative impact on the sales growth of indigenous firms. The volume of microeconometric studies of the relationship has increased since the 1990s, because much of the micro data at the firm- and establishment-levels has become available. For example, Kinoshita (2001), Girma (2005), and Todo (2008) showed that a positive impact of inward FDI on the growth of indigenous firms depends on whether or not indigenous firms have the ability to absorb technology.[6]

1.2.1.4 The Pessimism of Globalization
The pessimism over globalization and the negative influence of globalization have also been a long-standing topic of discussion. The related research branches on the relationship between globalization and growth includes the infant industry argument, structuralism, dependency theory, and so on. Although they are different on various counts, and the main subject of some of the studies is not technology, they implicitly share the perspective that the differences between developing and developed countries are due to insufficient technology. Among the lines of argument to examine pessimism, this book focuses on the infant industry argument because it has been developing for centuries.

An infant industry is defined as a nascent industry in developing countries. Although such an industry might be able to get a comparative advantage in the future, it must be temporarily protected for its low productivity in comparison with the mature industries in developed countries (Krugman et al., 2012).[7] The concept of the infant industry argument can be traced back to mercantilism in the seventeenth century

(Irwin, 1996). Mercantilism, including protective trade, was criticized by Adam Smith, but still the idea of protection has been maintained. When the Industrial Revolution took place in the United Kingdom (UK) in the late eighteenth century, latecomers, such as the US and Germany, realized that protective trade was necessary to deal with the first mover. At first, Hamilton (1791) highlighted the need for customs duties to develop the US' manufacturing sector.[8] This is because the UK's mature firms had significant advantages in both quality and price over the US firms. Also List (1841) explored the need for protective institutions that were necessary to develop Germany's manufacturing sector.

The infant industry argument is sophisticated in terms of its justification and criteria for protection. Its justifications have been underpinned by the lack of dynamic economies of scale and the lack of spillover externalities (Corden, 1997). Dynamic economies of scale indicate an effect that decreases the average cost along with increasing accumulated production volumes, the so-called experience effect or learning effect. The spillover externalities describe that private incentives probably do not create optimal social production if one's experiences also increase competitors' productivity. Government intervention can be justified only in cases of market failures due to externalities. This means that every nascent industry cannot be protected unconditionally.

In addition, the criteria for protection also have been tightened (Baldwin, 1969). The following criteria given by Mill, Bastable, and Kemp have to be satisfied. Mill's criterion states that the average cost of firms in developing countries should decrease compared to that in developed countries. Next, Bastable's criterion states that an increase in social benefit from protection in the end should exceed the decrease of the benefit at the beginning. These criteria are mere comparisons between the costs and benefits of protection; consequently, they do not concern the necessity of government intervention (Ito et al., 1988). On the other hand, Kemp's criterion relates to government intervention. Kemp (1960, 1964) criticized the widely used Mill–Bastable criteria and argued that intervention should be limited to the cases of market failure, specifically the existence of externalities in infant industries.[9]

There are other aspects to be considered. Sauré (2007) argued that protection does not necessarily promote the accumulation of new technologies in developing countries. This happens because indigenous firms are more likely to prefer technologies used in the past, even if there are new technologies that can enable a firm to increase its productivity. In contrast, arguments favoring protectionist have also been propounded. Baldwin (1969) and Corden (1997) showed that domestic policies of protection, such as subsidies, are better in terms of welfare distortions than trade policies, such as tariff modification. According to Melitz (2005), however, subsidies are not realistic for developing countries due to the burdens of expenditures that entail. He argued that quota systems are better than tariffs in terms of social welfare, because the former brings domestic prices close to international prices and increases LBD, whereas the latter does not. In this way, the infant industry argument has also been evolving in the context of globalization.

1.2.1.5 Research Question

Several studies have been conducted on the relationship between globalization and developing countries and their firms. These studies have reached a consensus that growth depends on technology diffusion and its conditions to realize diffusion, as described in many studies. In other words, growth depends on the equilibration of technological levels between developing and developed countries or between their firms.

However, an examination of the actual growth of indigenous firms in developing countries reveals that indigenous firms have been growing in spite of the lack of technology, as briefly shown in the Introduction. Although indigenous firms in China's electronics industry have faced a technology gap, they have offset the gap by using both external technology and internal knowledge. Therefore, in addition to the conditions of technology diffusion, there may be another condition required for the growth of indigenous firms.

Therefore, studies must be conducted on the phenomenon that indigenous firms can grow by diversifying the make-or-buy decision in comparison to that of foreign firms. To the effect of the make-or-buy decision, this book is closely related to Antràs and Helpman (2004) and Antràs (2005). These studies focused on the boundaries of firms in developed countries in the context of globalization. They analyzed the conditions under which firms in developed countries make products in-house or outsource them, demonstrating a multinationalization mechanism for firms. In this study, we analyze the boundaries of indigenous firms in developing countries in the context of globalization. In the growing research field on the boundaries of firms in a globalization regime (Helpman et al., 2008), we focus on the make-or-buy decision of indigenous firms under the influence of foreign firms, in order to comprehend the growth of indigenous firms in the context of globalization. The next section reviews the relationship between openness and the economic growth in China and shows that openness has facilitated technology diffusion and growth while diversifying the boundaries of indigenous Chinese firms.

1.2.2 CASE STUDY OF CHINA

1.2.2.1 Globalization and Growth
To verify the diversification of indigenous firms in developing countries, we study the growth of indigenous Chinese firms in the globalization regime. The Chinese economy has begun to grow rapidly since the economic transformation, the so-called economic reform and open-door policy (*gaige kaifang*), launched in 1978 (Kojima, 1997; Nakagane, 2002, 2010; Naughton, 1995, 2007; Wedeman, 2003; Wu, 2010). China has been growing relatively faster than other transition economies (Kimura, 2006a). The Chinese government

had closed the economy due to worsening ties with the superpowers, that is, the US and former Soviet Union, during the Cold War era. Subsequently, when China opened up to the world, its open-door economic policy made a significant impact on the nealy closed economy. The open-door policy is one of the two wheels that propelled economic transformation. The other is, of course, the economic reform. The open-door policy has stimulated international trade and inward FDI. And China was finally admitted into the World Trade Organization (WTO) in 2001. The outward FDI from China has also been increasing since the late 1990s, and today China is a major investor in the world. Consequently, openness has had greatly influenced the economic growth of China, so its relationship between openness and growth has been studied intensively (Branstetter and Lardy, 2008; Kojima and Horii, 2007; Marukawa et al., 2014; Ohashi, 2003; Pei et al., 2008).[10]

It has been recognized that the open-door policy has had a positive effect on the economic growth of China, and technologies have diffused by virtue of the economic liberalization (Fan, 1992; Jiang, 1993; Li, 2009).[11] Technologies were intensively introduced along with importing plants and equipment associated with factory construction until the mid-1990s (Kimura, 2013a). Introduction of technology through patents and designs is generally not adequate for developing countries of which the ability to absorb technology is poor and which face a significant technology gap (Yin, 2003). However, it is difficult to make high-quality products using only hardware-related technologies such as plants and equipment. So the knowledge necessary for manufacturing management has also been introduced (Hao, 1999; Yuan, 2009). Thus, technology introduction has expanded since the mid-1980s (Marukawa, 1990; Maruyama, 1988). In addition to plants and equipment, technology transfer and consulting services also have become major introduction channels. Consequently, the content of technology to be introduced has been upgraded along with the growth of Chinese firms and the increase in R&D investment. Therefore, it can be said that openness has contributed to growth through technology diffusion.

In addition, inward FDI from developed countries also has contributed to the diffusion of advanced technologies and management know-how to Chinese firms. MNE subsidiaries in China have introduced technologies from their home countries, and thus, technologies, which were initially not in China, spread and accumulated steadily. Also, inward FDI has created a knowledge spillover to Chinese firms indirectly. The Chinese government has actively sought to open the economy and to receive inward FDI; therefore, inward FDI has had a significant influence on economic growth and industrial development. Qiu (2009) analyzed the relationship between inward FDI and total factor productivity (TFP) at the industry-level and demonstrated a spillover effect through inward FDI. TFP is the productivity generated by the total input, not by individual inputs, such as capital and labor. It is generally considered a technological change, though it is merely a residual seen from the productivities of the individual inputs. Tuan et al. (2009) showed the positive impact of inward FDI on TFP in Pearl River Delta and Yangtze River Delta, where the manufacturing sector has been significantly developing and has become agglomerated. Jiang and Zhang (2006) also showed that the higher the technology absorption ability, the bigger the spillover effect. In other words, if technologies do not diffuse to industries in which the ability is significantly low. The channels of technology diffusion have also been studied. Ito et al. (2012) identified a forward linkage of the positive effect of foreign firms on the TFP of indigenous firms.

With the growing Chinese economy, a variety of industries have also been rapidly developing in China. In particular, labor-intensive industries have been significantly expanding owing to its abundant labor force.[12] During the planned economy era, the Chinese government had ignored China's comparative advantage and emphasized capital-intensive or heavy and chemical industries, such as steel, chemical, machinery industries, and so on (Lin et al., 1994). Subsequently in the economic liberalization regime, the resource allocation policy has been changed in line with the comparative advantage of China. Therefore, the economic growth in China can be explained on the basis of the comparative advantage in terms of international division of labor among countries and the growth of Chinese original equipment manufacturers (OEMs), which make products for foreign firms. However, it cannot fully explain the growth of indigenous Chinese firms with own brand names grew under fierce competition with foreign firms, particularly because foreign firms also enjoy the comparative advantage when investing in China or transacting with factories in China. That numerous manufactures operate successfully with their own brand names is a distinctive feature. Some of them in China's electronics industry are Haier Group (Haier), Hisense Group (Hisense), Sichuan Changhong Electric (Changhong), TCL, Konka Group (Konka), Lenovo, Gree Electric Appliances (Gree), Midea Group (Midea), Xiamen Overseas Chinese Electronic (XOCECO), Huawei Technologies (Huawei), ZTE, and so on. Against this backdrop, we investigate how indigenous firms with their own brand names have attained rapid growth against fierce competition from foreign firms.

1.2.2.2 Diversification of Organizations

In the fast-growing manufacturing sector, China's electronics industry has been developing significantly under globalization.[13] However, Chinese firms have not been completely homogenized with foreign firms through technology acquisition. They have been growing by outsourcing technologically difficult production stages and by focusing on the sales stage, as has already been discussed. This characteristic is shared with other manufacturing industries, especially the machinery industry, such as the automotive industry, the motorcycle industry, and so on (Marukawa, 2007;

Ohara, 2006; Watanabe, 2013). This growth pattern has been emphasized as a growth strategy for Chinese firms (Ling, 2005).

Therefore, many related features have been investigated in area studies on China. For example, Marukawa (1996, 2007) showed that Chinese television (TV) set manufacturers accumulated the know-how to use key components such as cathode-ray tubes (CRTs), produced by various suppliers. Because few Chinese TV set firms make CRTs by themselves, most buy them from indigenous specialized suppliers (Shanghai Caijing Daxue Ketizu, 2006). In addition, Ohara (1998, 2000) showed that indigenous air-conditioner manufacturers also do not make key components such as compressors, but they focus on organizing their sales and after-the-sale service networks. Kimura (2006b, 2010b) showed that major indigenous mobile phone handset firms do not design new models, but adopt the marketing-oriented strategy and focus on designing the appearance of handsets and developing nationwide sales networks.

These characteristics of the development and sales stages are closely associated with the development of the vertical specialization and the existence of the home advantage. First, the vertical specialization has been effectively developing in the global electronics industry. Therefore, firms often buy key components and services from outside specialized firms (Baldwin and Clark, 2000; Fujimoto and Shintaku, 2005; Yasumoto, 2010).[14] In particular, Breznitz and Murphree (2011) argued that Chinese firms can continue to grow without focusing on developing innovative products because the Chinese manufacturing sector is fiercely competitive in manufacturing and the domestic market is significantly large.

The change in industrial structures has been introduced by technological changes based on digitalization of products and modularization of product structures. Both these changes are related to each other, especially in the electronics industry. Generally digitalization facilitates modularization of product structures by allowing for easier control of product functions.[15] The personal computer (PC) industry is a case well known for its vertical specialization. Intel's chips and Microsoft's Windows have become the key components and software for PCs. Platforms also have become the key component for handsets (Imai and Shiu, 2008, 2011; Kimura, 2006b; Li, 2010; Shiu et al., 2008). Handset manufactures can efficiently design new models using platforms that support the core functions of handsets. Thus, the technological progress on each key component has ultimately increased the function and value of entire products.

Next, in the sales stage, indigenous and foreign firms generally have the home advantage and away disadvantage, respectively, since indigenous firms can efficiently comprehend domestic markets and conduct marketing activities than foreign firms. Foreign firms generally have the technology advantage since they started businesses earlier than indigenous firms. In contrast, indigenous firms can comprehend domestic markets and conduct marketing activities more efficiently than foreign firms. Moreover, the majority of indigenous firms' managers were born and raised in their own countries. Therefore, foreign firms do not have enough knowledge of foreign markets and do not have strong positions in foreign business networks (Johanson and Vahlne, 1977, 2009). Thus, the greater the difference there are among the markets, the more effective the home advantage can be.

In international economics, it is well known that only competitive firms that possess certain advantages in technology, brand names, and so on, can invest in foreign countries (Dunning and Lundan, 2008). MNEs need to bear some additional costs to invest in foreign countries. In other words, only productive firms can export and invest because foreign firms are required to collect related information and organize sales networks in unfamiliar foreign markets (Antràs and Helpman, 2004; Helpman et al., 2004; Melitz, 2003). This has been corroborated by many empirical studies (Kimura and Kiyota, 2006; Mayer and Ottaviano, 2008). This, of course, does not directly show the away disadvantage since indigenous firms also have to bear some additional costs to start businesses even in domestic markets. However, these additional costs are strongly related with human resources. Therefore, indigenous firms generally enjoy some advantages in these activities in comparison with foreign firms. Actually, indigenous firms have been expanding their businesses by selling products that can meet Chinese consumers' tastes (Liu, 2002). After all, it is difficult for foreign firms to comprehend the market (Björkstén and Hägglund, 2010).

As the Chinese market is large, rapidly growing, and diversified as well, it can be difficult for firms to adopt to fluctuating market trends and effectively sell products throughout the country.[16] China has the world's largest population at over 1.3 billion, which is widely dispersed geographically across the vast country. In addition, the Chinese market is diversified, especially in terms of income classes, such as the high-end, middle-end, and low-end markets (Kimura, 2006b). Therefore, the Chinese market includes a vast market for high-quality and multi-function products mainly in urban areas, and another vast one for price-competitive products with only basic functions, mainly in rural areas. However, there are many rural areas that are difficult to access (Tse, 2010). Consequently, understanding consumers' needs and preferences regarding design and functions within a short span of time in every market is a challenging task for foreign firms. It is well known that Haier has realized rapid growth by improving services that are strongly required by consumers (Mo, 2013). In addition, if firms try to expand their sales, they must acquire efficient channels connecting rural areas located at a distance from large cities. Of course, firms can choose to have distribution channels rooted in the planned economy era. However, traditional channels are often ineffective and have limited delivery capacity (Arai, 2005; Sekine, 2014). Moreover, it becomes difficult for firms to control detailed and lengthy distribution processes if all sales activities are left to outside firms (Kimura, 2006b). Therefore, firms must organize their own channels in partnership with across-the-country distributors and retailers so as to access local and rural markets, while specific marketing strategies vary from firm to firm. These features of sales are salient, especially in China, but every emerging economy also shares similar characteristics. Therefore, other emerging economies can draw a lesson by understanding the growth of indigenous Chinese firms.

Finally, it is noteworthy that protectionist policies also contributed to domestic industrial development under the gradual liberalization, especially until China's accession to the WTO. For instance, even though foreign firms entered the Chinese market, they faced some restrictions in selling in China because they were generally not allowed to sell products other than those made in their factories located in China. Rights for imports and domestic sales were limited; therefore, foreign firms could not increase their lineups of products, although joint ventures (JVs) with indigenous Chinese firms were conditionally allowed. Also, in production, investment was limited to specific industries and areas, and they were often required to affiliate with indigenous firms. The consequence was that Chinese firms were afforded ample room for growth. However, the growth of indigenous firms is not the sole cause of protectionist policies in the case of China. Although the market was protected to some extent, indigenous firms have been competing with other indigenous and foreign firms and have improved their competitiveness through fierce competition (Kimura, 2011a). We will discuss it in the next chapter.

1.2.2.3 Research Question

Although previous studies showed that Chinese electronics firms have realized rapid growth by using the development of the vertical specialization and the home advantage aggressively, to the best of the author's knowledge, the diversification in organization has not been systematically investigated. Since the relationship between indigenous and foreign firms has not been considered, the rational selection of indigenous firms to identify the optimal balance between integration and disintegration of business stages has not been explicitly analyzed. If we do not incorporate the existence of foreign firms in the boundary selection of indigenous firms, then we cannot evaluate the extent to which indigenous firms integrate or disintegrate each stage of value chains. It can be hardly expected that they buy all of the development services from outside firms, and similarly integrate all of the marketing services. Stated differently, it must be irrational either to disintegrate without end or to integrate without end. Therefore, to determine rationality, we analyze the make-or-buy decision of indigenous firms in the context of competition with foreign firms.

1.3 FRAMEWORK AND METHODS

To illustrate a diversification mechanism for indigenous firms facing a technology gap, we need to show that (1) there is a technology gap between indigenous and foreign firms and (2) indigenous firms offset the technology gap by differentiating their boundaries. The former is the precondition to argue the latter. If technologies completely diffuse from developed countries to developing ones, we may not be able to find such a contrast in organization between indigenous and foreign firms. However, it is hard to imagine that technologies completely diffuse, or that technologies never diffuse. As the reality lies somewhere in between, we investigate the behavior of indigenous firms with incomplete technology diffusion.

To show that there is a technology gap, this study applies two methods to illustrate the technology gap. First is an empirical study on the relationship between inward FDI and TFP. When examining the relationship between inward FDI and the productivity of indigenous firms, if inward FDI increases productivity, then inward FDI will possibly foster technology diffusion. However, if inward FDI does not increase productivity, then inward FDI will not possibly facilitate technology diffusion. The relationship between inward

FDI and productivity is analyzed using a method based on previous studies. It is common among the previous studies that the levels or growth rates of production or productivity of firms were usually regressed on variables that show the presence of foreign firms (specifically, the ratio of foreign firms' capital to the total capital in an industry). Therefore, the spillover or market-stealing effect can be found depending on whether the parameters of this variable are positive or negative. Based on this study, we show that technologies do not always diffuse completely.

Another method is a case study of technological hurdles in electronics products. We examine the nature of a technology gap. In the existing literature on technology diffusion, it has been generally assumed that a product embodies a single technology, whereas in reality, several technologies are typically required as part of an entirely hierarchical system of technologies to finish a product. Firms that engage in product differentiation so as to increase the value of products are required to develop advanced technologies in comparison with standardized technologies that are used generally among all firms in the same market. However, to develop advanced technologies and to propose ideas that are critical to product differentiation on the basis of advanced technologies, firms first need to have sufficient experiences and know-how in the use of the lower-level technologies of advanced ones. Higher-level technologies, that is, advanced technologies, cannot work in isolation from the lower-level technologies. In other words, the knowledge of the lower-level technologies can very well be the base for, and can accurately drive, advanced technologies. This implies that latecomers or indigenous firms in developing countries will be at a disadvantage because they have little experience in product development, which is needed to absorb all of a technology in a short span of time. Moreover, know-hows that have accumulated through LBD cannot be diffused from learners. This feature is shared through the learning curve or experience curve effect for decreasing unit cost with increasing cumulative production volume. The existing cumulative volume of foreign firms is already large and therefore their unit cost is sufficiently low. In contrast, the cumulative production volume of indigenous firms is still small. Therefore, their unit cost is also not yet low. Thus, we can find the technology gap by focusing on the technological structures of products and their complexities.

To study the diversification mechanism, a model of the boundary selection of indigenous firms facing a technology gap has been analyzed in this book. We integrate the existence of the technology gap into the concept of the boundaries of the firm. The model of the boundaries was developed by Grossman and Hart (1986) and Hart and Moore (1990) (the GHM model).[17] The boundaries of the firm are determined to resolve the hold-up problem incurred by incomplete contracts in the GHM model.[18] To understand the problem, assume there is a manufacturer and a supplier, and that the value of the buyer's products can be increased by investing in human capital of the seller only for the buyer's products. Human capital includes factors such as technology skills, knowledge, and accumulated experiences, a high level of each of which can increase the value of products and enhance the firms' profits. The seller invests in human capital and makes intermediate goods or services to increase the value of the buyer's products. The contract, however, is incomplete, because the firms cannot draw up all the unforeseen contingencies in advance. Therefore, incomplete contracts cannot ensure that the buyer would inevitably buy intermediate goods or services made by the seller. As a result, the seller will hesitate in agreeing to such a relationship-specific investment and will be concerned about a sunk investment due to contractual infirmity. The risks associated with the agreement can result in underinvesting in human capital for both parties. This inefficiency in investment for mutual benefits is called the hold-up problem. To solve this predicament, the buyer can decide to independently produce the components by themselves to maximize the value of products through human capital investment. Consequently, the hold-up problem caused by incomplete contracts serves as the basis for internalized boundaries.

However, the model does not deal with the technological level of firms facing the boundary selection. Therefore, we incorporate the existence of the technology gap, that is, the lack of LBD, into the boundary selection, explicitly analyzing the boundary selection of indigenous firms in developing countries. Thus, the decision of indigenous firms can be changed as follows. Even if integration is rational according to the theory of the boundaries of the firm, when indigenous firms face the significant technology gap, they are unable integrate the production stage related to the technology. Instead, they choose to buy goods and services from outside firms instead of integration. This is because even if they integrate the stage and invest human capital in the stage, the investment cannot increase the product value because they do not have enough experiences in the use of the technology. As a result, the investment can amount to nothing. In contrast, if there are any stages in which indigenous firms can invest more effectively than foreign firms in order to compensate for the technology gap, then indigenous firms would integrate the stage while foreign firms might not do it. Hence, the boundary selection depends on the technology gap as well as the rationality according to the theory of the boundaries of the firm.

1.4 CONCLUSION

In this chapter, the background and the framework of this study have been explained. Based on them, we analyze Chinese electronics firms and elucidate a diversification mechanism. To do this, we particularly focus on three factors: the technology gap, external technology, and internal knowledge. By analyzing these three factors, we find a third way of the growth of indigenous firms driven by the diversification mechanism, although previous studies have focused on the positive or negative influence of globalization.

CHAPTER 2

OUTLOOK

2.1 INTRODUCTION

This chapter first reviews the object of this book—development of China's electronics industry in the midst of globalization. The study aims to elucidate the growth mechanism of indigenous firms under the globalization regime. Taking up China's electronics industry as a representative case to be studied in this context turns out to be correct as we go on to elucidate (1) the growth of Chinese electronics firms in order to understand the growth mechanism of indigenous firms, (2) the competition with foreign firms so as to understand the influence of globalization, and (3) the development of the vertical specialization in the industry in order to examine the various make-or-buy choices of indigenous firms.

Before beginning the discussion, let us first define the electronics industry. As stated in the Introduction, the electronics industry covered in this book includes consumer electronics products, home appliances, and so on, some of which are usually included in the electrical industry. We separate the electrical and electronics industries when data is classified on the basis of industrial classifications. Specifically, according to the Industrial Classification for National Economic Activities of China, the electrical and electronics industries are included in the categories "Manufacture of Electrical Machinery and

Apparatus" and "Manufacture of Computers, Communication and Other Electronic Equipment," respectively. And, according to the Harmonized Commodity Description and Coding System (Harmonized System: HS) for international trade, the electrical and electronics industries are generally included in the categories "Nuclear reactors, boilers, machinery and mechanical appliances; parts thereof," and "Electrical machinery and equipment and parts thereof; sound recorders and reproducers, television image and sound recorders and reproducers, and parts and accessories of such articles," respectively.[1]

In this chapter, we review items (1) through (3) of the foregoing list, one in each section, and then we conclude the chapter.

2.2 Presence

The electronics industry has been a leading one in the fast-growing manufacturing sector in China (Kimura, 2007a). The secondary industry, including the manufacturing sector, accounts for roughly half of the rapidly increasing GDP of China (Figure 2.1). As a result, China came to be called "the world's factory" in the early 2000s. The electrical and electronics industries have strengthened their presence in the vital manufacturing sector. The electrical and electronics industries have accounted for more than 14 percent of the total gross industrial output value, and more than 10 percent of the total value added (Figures 2.2 (a) and (b)). Figure 2.2 tells us that the industries have developed during the 1990s and 2000s. Of course, some of the electronics products are made by foreign firms with production units in China. Actually, 80 percent of China's exports from the information and communication technology (ICT) industry were accounted for by foreign firms in 2009 (*Zhongguo Xinxi Chanye Nianjian Bianweihui*, 2010). However, it is noteworthy that not only foreign firms but also a large number of indigenous Chinese firms have been growing rapidly.

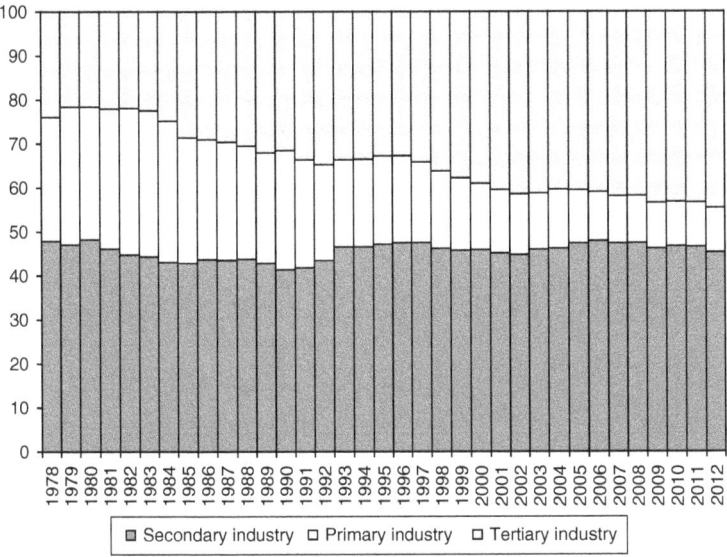

Figure 2.1 Composition of GDP by industry, 1978–2012 (%)
Notes 1: To show the share of the secondary industry clearly, the industry is put at the bottom of the figure.
2: According to National Bureau of Statistics of China (2013), the primary industry includes the agriculture, forestry, animal husbandry and fishery industries. The secondary industry includes the mining and quarrying, manufacturing, production and supply of electricity, water and gas, and construction industries. The tertiary industry includes all other economic activities.
Source: Author's creation based on revised data from National Bureau of Statistics of China (2013).

With the growth in production volume, the domestic market also has been expanding throughout the transition era. A variety of products have been diffusing into the urban market with time (the upper part of Table 2.1). Following the lead in the urban market, the rural market also has started expanding rapidly since the 1990s (the lower part of Table 2.1). As a result, any product is always contributing to the expansion of the Chinese market. Although the income disparity is, of course, a major concern in China, an important consideration in examining the growth of indigenous Chinese firms is the fact that the rapid expansion of the domestic market is consistent.

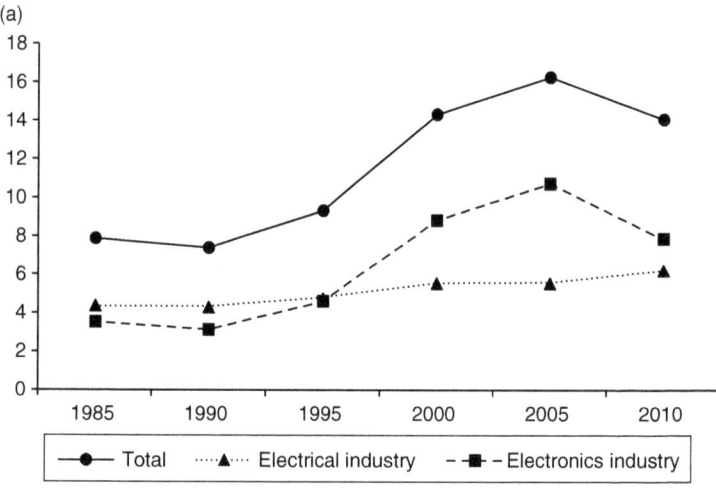

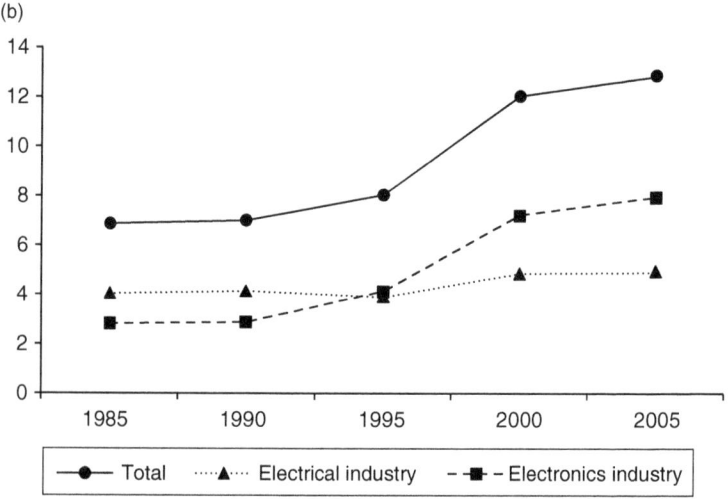

Figure 2.2 Market presence of the electrical and electronics industries (a) The ratio of the gross industrial output value of the electrical and electronics industries to the total industrial output value, 1985–2010 (%). (b) The ratio of value addition in the electrical and electronics industries to the total value added of all industries, 1985–2005 (%)

Source: Author's creation based on revised data from National Bureau of Statistics of China (various years).

Table 2.1 Penetration rates, 1980–2012 (%)

Item	1980	1985	1990	1995	2000	2005	2010	2012
Urban households								
Washing machine	n.a.	48.3	78.4	89.0	90.5	95.5	96.9	98.0
Refrigerator	n.a.	6.6	42.3	66.2	80.1	90.7	96.6	98.5
Black & white TV set	}32.3	66.9	52.4	28.0	n.a.	n.a.	n.a.	n.a.
Color TV set		17.2	59.0	89.8	116.6	134.8	137.4	136.1
Air conditioner	n.a.	n.a.	0.3	8.1	30.8	80.7	112.1	126.8
Mobile-phone handset	n.a.	n.a.	n.a.	n.a.	19.5	137.0	188.9	212.6
PC	2.8	n.a.	n.a.	n.a.	9.7	41.5	71.2	87.0
Camera	n.a.	8.5	19.2	30.6	38.4	46.9	43.7	46.4
Video camera	n.a.	n.a.	n.a.	n.a.	1.3	4.3	8.2	10.0
Microwave oven	n.a.	n.a.	n.a.	n.a.	17.6	47.6	59.0	62.2
Rural households								
Washing machine	n.a.	1.9	9.1	16.9	28.6	40.2	57.3	67.2
Refrigerator	n.a.	0.1	1.2	5.2	12.3	20.1	45.2	67.3
Black & white TV set	}0.4	10.9	39.7	63.8	53.0	21.8	6.4	1.4
Color TV set		0.8	4.7	16.9	48.7	84.1	111.8	116.9
Air conditioner	n.a.	n.a.	n.a.	0.2	1.3	6.4	16.0	25.4
Mobile-phone handset	n.a.	n.a.	n.a.	n.a.	4.3	50.2	136.5	197.8
PC	n.a.	n.a.	n.a.	n.a.	0.5	2.1	10.4	21.4
Camera	n.a.	n.a.	0.7	1.4	3.1	4.0	5.2	5.2

Source: Author's creation based on revised data from National Bureau of Statistics of China (various years).

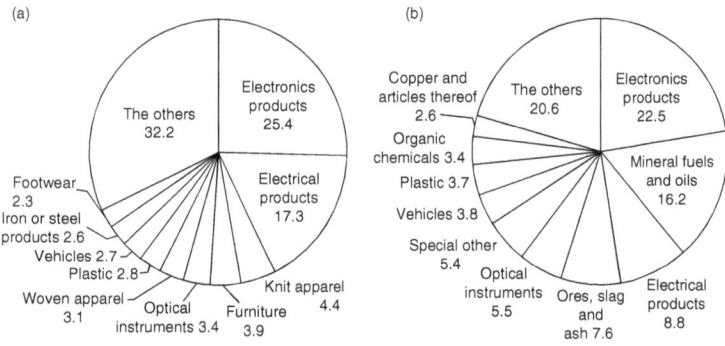

Figure 2.3 Exports and imports, 2013 (%) (a) Exports (b) Imports
Source: Author's creation based on revised data from World Trade Atlas provide by Global Trade Information Services.

In addition to the expansion of the domestic market, the exports of electrical and electronics products are also increasing. The electronics and electrical industries are the first and the second largest exporters, respectively (Figure 2.3). This trend is not only the latest but it was there throughout the 2000s, although we cannot ignore the fact that foreign firms account for a large portion of the exports. Moreover, the industries have recorded trade surpluses since the mid-2000s and have contributed significantly to foreign currency earnings. However, because the industries had imported huge quantities of key components and production equipment in comparison with exports of electrical and electronics products, both the industries had trade deficits before the mid-2000s. This is continuing because the imports of key components and production equipment increase as a result of larger exports. Since this shows a weakness of the industry in the international division of labor, we will discuss this later. The development of the industry is closely linked to the vigor of trade under globalization, in addition to foreign direct investment (FDI), as discussed in the next section.

2.3 COMPETITION WITH FOREIGN FIRMS

A characterizing feature of the economic growth in China is the absorption of a huge amount of FDI. Inward FDI has been increasing sharply since the 1990s, while foreign loans were more prevalent in the 1980s (Figure 2.4). In the 1990s, the electronics industry absorbed a large portion of the burgeoning inward FDI (Kimura, 2012a). Consequently, as shown in Figure 2.5, the number of foreign firms (FE in the figure) in both electrical and electronics industries (Manufacture of Electrical Machinery and Apparatus and Manufacture of Computers, and Communication and Other Electronic Equipment, respectively, in the figure) is larger than that of indigenous firms (SOE and PE in the figure). Although many foreign firms have set up factories in China as bases for their export operations, they also aim to capture the fast-growing Chinese market for their further growth.

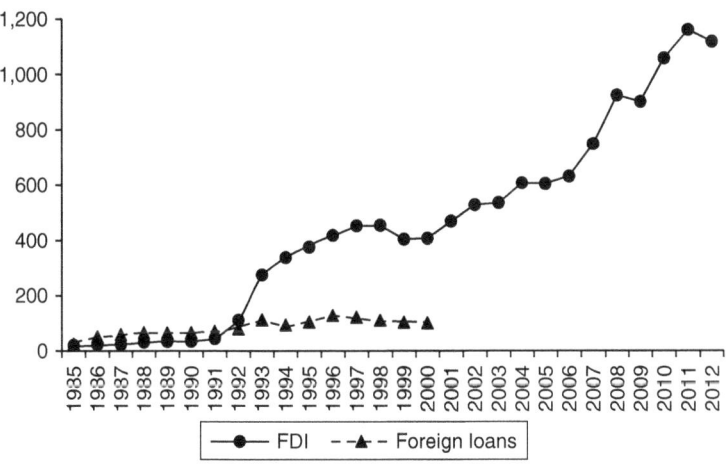

Figure 2.4 Inward FDI and foreign loans (Flow), 1985–2012 (100 million USD)

Note: "Foreign loans" are not documented in the statistical yearbooks after 2001.
Source: Author's creation based on revised data from National Bureau of Statistics of China (2013).

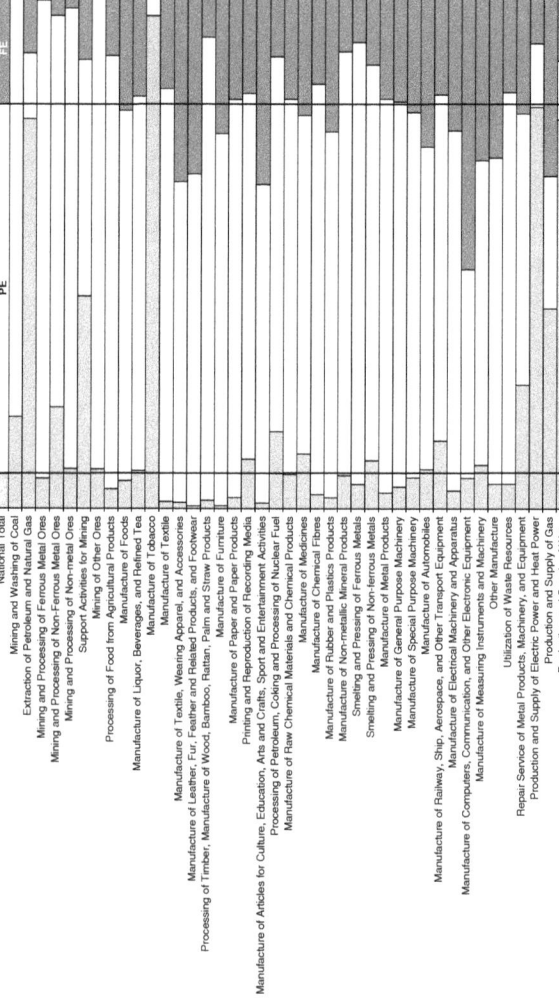

Figure 2.5 Proportion of firms by ownership, 2012 (%)

Notes 1: SOE, PE, and FE stand for state-owned enterprise, private enterprise, and foreign enterprise, respectively.
2: The vertical lines drawn from the boundaries of SOE, PE, and FE in National Total indicate the averages of the proportions by ownership.
Source: Author's creation based on revised data from National Bureau of Statistics of China (2013).

However, indigenous firms have dominated the domestic markets of various electronics products against foreign firms, while, before indigenous firms realized a rapid growth in the 1990s, foreign firms had a higher market share than what they have at present. Therefore, in this book we discuss the growth process of indigenous firms in terms of the organizational form, but competition is also a significant key factor for the rapid growth of indigenous firms. Indigenous firms have realized growth by developing competitiveness through fierce competition with other indigenous firms as well as foreign firms. Although competition with other indigenous firms is not the primary topic of this book, it has contributed to improving indigenous firms' competitiveness at the sales stage, which we focus on in this book. Therefore, we review the entry process of many indigenous firms and its effect on competition with foreign firms in the rest of this section, doubling as the introduction to the development of China's electronics industry.

China's electronics industry has been developing rapidly since the start of economic liberalization, although it was controlled to produce equipment for the defense industry in the planned economy era (*Zhongguo Dianzi Gongye 50 nian* Bianweihui, 1999). The Chinese government aimed to use the new technologies of electronics at that time, in order to be ready for future wars. The government placed a disproportionate emphasis on the production of military electronics goods, such as specialized telecommunication equipment, aviation navigation equipment, radar apparatus, and so on, rather than the production of consumer electronics products, such as television (TV) sets, refrigerators, and so on (Dong and Wu, 2004). Therefore, almost all of the major electronics firms had made defense goods. Numerous manufacturers were established, especially in the First Five-Year Plan period (1953–1957), with the support from the former Soviet Union, and during the Cultural Revolution (1966–1976) (*Dangdai Zhongguo* Congshu Bianji Bu, 1987).

The government, however, changed the military goods-centered policy of the electronics industry in the late 1970s. Although the government established many military goods manufacturers, the demand for such products decreased when the signs of global tension eased in China through improved relationships with the Western countries. Consequently, the government altered the policy and began encouraging consumer goods production to survive the decreased demand. For example, Sichuan Changhong Electric (Changhong) founded in 1958 had produced military radar apparatus, but it began producing black-and-white TV sets in 1972 and color TV sets in 1985 by adopting production lines from Matsushita Electric Industrial (currently Panasonic, Japan). Therefore, the first entrants in the industry were manufacturers that went through an operational change, that is, from producing military equipment to consumer goods. As a result, the ratio of the value added to consumer goods to the value added to military equipment reached 97.0 percent in 1991, which was mere 20.0 percent in 1979 (Komagata, 2000).

Not only manufacturers originated in the defense industry but also many new entrants started businesses in the industry. At the beginning, numerous state-owned enterprises (SOEs) and township-and-village enterprises (TVEs) entered the industry. For example, Haier Group (Haier) is an SOE controlled by the Qingdao local government. And Gree Electric Appliances (Gree) is a TVE controlled by the Zhuhai local government. Although a lot of indigenous firms entered the electronics industry, almost all of them were firms controlled by local governments, they were not private firms. Therefore, some say that the economy has been controlled by the government. Of course, only local governments had enough money to establish firms at that time. However, entrants had faced fierce competition even among SOEs and TVEs in the electronics industry. To emphasize, firms have been competing with each other since the 1980s, making Chinese firms competitive as a whole. In addition to firms established by local governments, many private and foreign firms also entered the fray in the 1990s when the economy was much more liberalized in the early 1990s. Because the Chinese market has been expanding rapidly, a lot of entrants have been able to find room for entry and survive in the market.

Thus, indigenous firms have been growing and improving their market position through fierce competition with other indigenous firms as well as foreign firms. Numerous indigenous firms have entered almost all of markets by adopting strategies to decrease a variety of entry costs (Watanabe, 2013). Therefore, market structures have thus far been dispersive (Watanabe and Kimura, 2012), facilitating the rise of many major indigenous firms. Foreign firms enjoyed popularity, especially in the 1980s and the 1990s; however, products made by indigenous firms succeeded to gain much popularity among Chinese consumers in the 1990s because they improved their product quality. In many markets, indigenous firms have gained the highest market precense. Moreover, some of the major Chinese firms have focused on creating networks to increase nationwide sales and develop after-the-sale service. In addition, the fierce competition has forced each indigenous firm to make its sales network more effective than that of other competing indigenous firms that have been sharing the home advantage.[2] Competition has become one of the reasons of the rapid growth of indigenous firms.

In this way, a fierce competition has become salient to the electronics industry in China. We will explicate it in comparison with the case of India (Kimura, 2011a). Although China's and India's electronics industries initially had similar initial conditions, the present results significantly differ. Both in China and India the electronics industry was established with the aim of catering to the defense industry. With the partial economic liberalization since the 1980s, India's electronics industry had been expected to be a leading one (Joseph,

2004). However, the number of firms entering the Indian market was low in comparison with China, because in India the licensing system for entry was extremely rigid and prolonged recession impeded the market's growth. Consequently, the Indian market had been monopolized by a few major firms until the late 1990s. After India opened its market to the world in the 1990s, Indian indigenous firms have faced fierce competition from foreign firms, especially those from South Korea (Korea).

As stated in the previous chapter, the protection of the domestic market also partially contributed to the growth of indigenous Chinese firms in addition to the fierce competition. Protectionist policies facilitated a run-up period for indigenous firms to brush up their competitiveness. However, protection alone could not explain the growth of indigenous firms. Although indigenous Indian firms also had the protected run-up period until the late 1990s, the market became dominated by Korean firms since the open-door policy (Kimura, 2011a). Indigenous firms in countries that adopted the strategy of import-substitution industrialization (ISI) in the past had also lost their market share since the start of economic liberalization. Although there are a lot of indigenous electronics firms in many countries, the number of global firms is limited. Comparing Chinese firms with other indigenous firms, we find that competition has had a significant role for the growth of Chinese firms. Chinese firms, which have been improving their market position through fierce competition with other indigenous firms, have been expanding business although foreign firms too have been targeting the vast Chinese market.

2.4 Vertical Specialization

As the development of vertical specialization changed the product structures, as described in Chapter 1, transactions within the industry increased. Table 2.2 shows the input coefficients of intra-industry transactions in each industry.[3]

Input coefficient is the input-to-output ratio. The input coefficient of intra-industry transactions in "Manufacture of Machinery and Equipment," which includes the electronics industry, is higher along with the coefficients in "Manufacture of Textile," "Wearing Apparel and Leather Products," and "Chemical Industry" in comparison with that in the other industries. Moreover, the input coefficients of intra-industry transactions in the three industries account for nearly half of those of total intermediate inputs. We can find the possibility of the development of the vertical specialization in the industries from Table 2.2.

As a result, the development of vertical specialization has increased the indigenous firms' chances of entry and growth. Although firms had to learn entire technological systems before the development of the vertical specialization, they have got to be able to produce products when they buy key components from outside specialized firms. As production has been fragmented, production stages in which indigenous firms in developing countries can enter have also been expanding (Ando and Kimura, 2005; Hiratsuka and Kimura, 2008). The technological entry barrier has dramatically decreased. Consequently, there exist a variety of the make-or-buy decisions on well-fragmented value chains.

Although Chinese firms have been absorbing technologies from foreign firms and have been increasing research and development (R&D) investment in the technology expenditure, they have been facing the technology gap in contrast to foreign firms in the same market. As a result, China still has disadvantages in the international division of labor within the electronics industry. Although China has been exporting large amount of final goods, it has been importing large amount of expensive key components and production equipment as well. Tables 2.3 (a) and (b) show corresponding cases in the electrical and electronics industries, respectively. The trade specialization coefficient indicates the ratio of exports to total international trade in an industry. In specific terms, if the coefficient is close to 1, the

Table 2.2 Input coefficients of input–output table, 2010

	Agriculture, Forestry, Animal Husbandry and Fishery	Mining	Manufacture of Foods, Beverage and Tobacco	Manufacture of Textile, Wearing Apparel and Leather Products	Other Manufacture	Production and Supply of Electric Power, Heat Power and Water	Coking, Gas and Processing of Petroleum	Chemical Industry	Manufacture of Nonmetallic Mineral Products	Manufacture and Processing of Metals and Metal Products	Manufacture of Machinery and Equipment	Construction	Transport, Storage, Post, Information Transmission, Computer Services and Software	Wholesale and Retail Trades, Hotels and Catering Services	Real Estate, Leasing and Business Services	Financial Intermediation	Other Services
Total Intermediate inputs	0.4153	0.5479	0.7892	0.8000	0.7327	0.7443	0.8016	0.8064	0.7805	0.8199	0.8179	0.7395	0.5719	0.4030	0.4110	0.3502	0.4499
Intra-industry transactions	0.1330	0.1159	0.2066	0.4360	0.2858	0.3287	0.0604	0.4102	0.1910	0.3363	0.4099	0.0106	0.0899	0.0246	0.0525	0.0803	0.0390
Total value-added	0.5847	0.4521	0.2108	0.2000	0.2673	0.2557	0.1984	0.1936	0.2195	0.1801	0.1821	0.2605	0.4281	0.5970	0.5890	0.6498	0.5501

Source: Author's creation based on revised data from National Bureau of Statistics of China (2013).

industry is close to specialization in exports. In contrast, if it is close to −1, the industry is close to specialization in imports.

Tables 2.3 (a) and (b) present the best ten and worst ten items in terms of specialization in exports in the electrical and electronics industries. The two industries share similarities in that the final goods tend to be included in the top ten but components and production equipment tend to be included in the worst ten. For example, final goods, such as washing machines, air conditioners, TV sets, and so on, are on the upper sides in the tables. On the other hand, components and production equipment, such as machining centers, semiconductor devices, integrated circuits (ICs), and so on, are on the lower sides in the tables. Compared to assembling of final goods, the manufacturing of key components and production equipment is relatively less developed at present. Comparatively speaking, it tells that not only indigenous firms but also foreign firms do not produce as much components and equipment in China as the amount of final goods produced. As manufacturers produce final goods, the imports of core components and production equipment also increase.

2.5 Conclusion

In this chapter, we have shown the preconditions required to investigate the growth of Chinese electronics firms in globalization. Specifically, we have described the market presence of the electronics industry in China, competition with foreign firms, and the development of vertical specialization. China's electronics industry has attained significant development, especially in the large manufacturing sector. As a result, it has become the largest exporter in China. However, it has been importing many expensive key components and production equipment to assemble final goods. Of course, we cannot directly apply features of the industry level to the firm level. In addition, there are many manufacturers of key

Table 2.3 Trade specialization coefficients, 1995–2013
(a) The electronics industry

HS	Description	1995	2000	2005	2010	2013
84	Total	−0.52	−0.12	0.22	0.29	0.38
8404	Auxiliary plant	−0.78	−0.56	−0.34	0.86	0.96
8450	Washing machines	0.21	0.72	0.92	0.92	0.93
8402	Other steam generating boilers	−0.83	−0.57	−0.06	0.87	0.92
8415	Air conditioners	−0.45	0.58	0.84	0.85	0.89
8410	Hydraulic turbines	−0.51	−0.35	−0.41	0.75	0.89
8467	Hand tool	−0.33	0.44	0.88	0.86	0.87
8469	Typewriters and word-processing machines	0.76	0.55	0.78	0.91	0.86
8437	Machines for cleaning, sorting seed	−0.58	0.08	−0.09	0.72	0.85
8470	Calculating machines	0.86	0.83	0.76	0.80	0.85
8423	Weighing machines	0.31	0.69	0.78	0.80	0.84
8461	Machine tools for shaping, slotting, gear cut, etc.	−0.86	−0.65	−0.52	−0.50	−0.45
8446	Weaving machines	−0.96	−0.95	−0.92	−0.72	−0.47
8479	Other machines with individual functions	−0.91	−0.92	−0.81	−0.63	−0.49
8444	Machines for extruding, drawing, texturing or cutting man-made textile materials	−0.99	−0.87	−0.83	−0.53	−0.62
8460	Machine tools for honing or finishing metal, etc.	−0.77	−0.66	−0.80	−0.71	−0.68
8486	Machines of a kind used solely or principally for semiconductor boules or wafers, etc.	n.a.	n.a.	n.a.	−0.84	−0.75
8420	Calendering or other rolling machines	−0.84	−0.53	−0.85	−0.84	−0.77
8401	Nuclear reactors	−1.00	−1.00	−1.00	−0.41	−0.78
8478	Machinery for tobacco preparation	−0.98	−0.89	−0.70	−0.90	−0.79
8457	Machining centers	−0.97	−0.97	−0.98	−0.96	−0.91

Source: Author's creation based on revised data from World Trade Atlas provide by Global Trade Information Services.

Table 2.3 (b) The electronics industry

HS	Description	1995	2000	2005	2010	2013
85	Total	−0.01	−0.05	−0.01	0.11	0.12
8513	Portable electric lamps	0.87	0.98	0.97	0.98	0.98
8528	TV sets	0.41	0.91	0.97	0.94	0.97
8521	Video recording or reproducing apparatus	0.05	0.97	0.97	0.95	0.97
8508	Vacuum cleaners	0.74	0.89	n.a.	0.96	0.95
8519	Turntables, record and cassette players	0.91	0.98	0.95	0.92	0.94
8516	Electrical instantaneous or storage water heaters	0.62	0.88	0.92	0.92	0.92
8509	Electro-mechanical domestic appliances	0.80	0.94	0.95	0.94	0.91
8527	Reception apparatus for radio-telephony or radio-broadcasting	0.87	0.97	0.92	0.87	0.89
8545	Carbon electroïces, etc	0.37	0.22	0.53	0.69	0.79
8539	Electric filament or discharge lamps	0.66	0.67	0.31	0.59	0.74
8534	Printed circuits	0.02	−0.06	−0.10	−0.04	−0.01
8541	Semi-conductor devices	−0.11	−0.40	−0.45	0.18	−0.02
8536	Electrical apparatus for switching, ≤ 1000v	0.26	−0.09	−0.21	−0.16	−0.07
8538	Parts (8535-8537)	−0.84	−0.63	−0.50	−0.26	−0.11
8533	Electrical resistors	0.16	−0.41	−0.41	−0.25	−0.17
8547	Insulating fittings	−0.43	−0.03	−0.07	−0.23	−0.18
8532	Electrical capacitors	−0.08	−0.35	−0.49	−0.44	−0.20
8548	Waste and scrap of primary cells and batteries	−0.35	−0.24	0.28	−0.41	−0.34
8514	Industrial electric furnaces	−0.93	−0.90	−0.78	−0.65	−0.40
8542	Integrated circuits	−0.69	−0.65	−0.70	−0.68	−0.45

Source: Author's creation based on revised data from World Trade Atlas provide by Global Trade Information Services.

components and production equipment in China. However, the large amount of their import reflects the relative lack of technology in comparison with the production of final goods. Next we discuss the technology gap at the firm level in Chapter 2.

CHAPTER 3

TECHNOLOGY GAP

3.1 INTRODUCTION

This chapter verifies that a technology gap between indigenous and foreign firms exists.[1] As stated in the previous chapters, if technologies diffuse from developed countries to developing ones through international economic activities, then globalization will make a positive impact on developing countries and their firms. On the other hand, if technologies do not diffuse, then globalization will make a negative impact on them. By showing the relationship between globalization and its negative impact, we conclude that technologies do not always diffuse in globalization and that there might be a technology gap between indigenous and foreign firms.

Although some of the technologies have diffused, as shown in Chapter 1, Chinese electronics manufacturers have reported low productivity.[2] Ito et al. (2008) showed that the total factor productivity (TFP) growth rate of the electronics industry was 5.2 percent in Japan, 11.1 percent in South Korea (Korea), but only 2.8 percent in China during the period between 1999 and 2004, although some industries in China, such as the motorcycle industry, recorded higher growth rates during this period than the same industries in Japan and Korea. Hu and Jefferson (2002) showed that the influence of inward foreign direct investment (FDI) on the electronics industry was negative, although the influence of FDI on the textile industry was ambiguous. Jiang and Zhang (2006) also showed that the electronics industry received a negative impact in comparison with other industries. Motohashi and Yuan (2010) investigated the vertical path of technology diffusion from foreign assemblers with research and development (R&D) activities to indigenous suppliers in the electronics and automotive industries. They did not find technology diffusion in the electronics industry, although they found it in the automotive industry. In this way, technologies do not always diffuse from foreign firms into Chinese electronics firms. Depending on industries, the influence of globalization is either ambiguous or clearly discernible.

Therefore, we investigate in detail the influence of globalization on China's electronics industry. At first, we divide the industry into 38 sectors and identify the characteristics of each sector. To investigate why the results of previous studies are ambiguous, differences among industries must be considered. A national economy comprises various industries and each industry depends on different technologies and know-how. In fact, although the overall TFP in China has increased since the economic reform, there are vast differences among industries. We identify the levels of TFP and value added as the performances of Chinese electronics firms. Therefore, value added and TFP at the firm level are used as proxies for the growth of the firm.

Second, we estimate the relationship between the growth of indigenous firms and the amount of inward FDI in each sector in this industry. The ratio of fixed assets owned by foreign firms to the total fixed assets in China is used as a proxy for inward FDI. In this chapter, we examine the relationship between the positive or negative effect of inward FDI and the technology gap between indigenous and foreign firms in every sector.

On the basis of this examination, we link the characteristics of each industry and the results of the impacts, and reveal that sectors with large technology gaps resulting from the lack of business experience in indigenous firms tend to experience negative effects from inward FDI. By comparing every sector in the industry, this chapter elucidates the influence of inward FDI on each sector.

This chapter is organized as follows. In Section 3.2, we introduce our dataset and estimate productivity using firm-level production functions. In Section 3.3, we estimate the effects of inward FDI on the growth of indigenous firms and discuss the results related to the findings in Section 3.2. Section 3.4 concludes the chapter

3.2 Dataset and Estimation of Productivity

In this section, we introduce the dataset and variables employed in the present analysis. And using the dataset, we estimate the productivity of indigenous and foreign firms.

3.2.1 Dataset

The dataset employed in this study is extracted from the ORIANA which is a database of company accounting information developed by Bureau van Dijk, a major provider of business information. The ORIANA database includes information from 2003 to 2007 for approximately 6.3 million indigenous and foreign firms in Asia, of which approximately 300,000 firms are located in China. Although there are an estimated five million firms in China that are required to file financial statements annually with the Chinese government, ORIANA includes approximately 300,000 Chinese firms because of limitations in obtaining information on all firms in China.

In our analysis, we use the ORIANA data on 28,948 firms in China in the electronics industry. Of the 28,948 firms in the dataset, 20,557 are indigenous firms and 8,391 are foreign firms. In this chapter, indigenous firms are defined as only firms that are wholly owned locally. Foreign firms refer to firms that are wholly owned or partly owned by foreign capital. Therefore, joint venture (JV) companies are also regarded as foreign firms in this chapter.[3]

We classify the electronics industry into 38 sectors using the six-digit North American Industry Classification System (NAICS) classification code (Table 3.1). Technologies for the industry are closely grouped at the product level. In NAICS, the electronics industry includes codes beginning

Table 3.1 List of the 38 sectors in the electronics industry

Five-digit	Six-digit	Definition
33341		Ventilation, Heating, Air-Conditioning, and Commercial Refrigeration Equipment Manufacturing
	333412	Industrial and Commercial Fan and Blower Manufacturing
	333415	Air-Conditioning and Warm Air Heating Equipment and Commercial and Industrial Refrigeration Equipment Manufacturing
33411		Computer and Peripheral Equipment Manufacturing
	334111	Electronic Computer Manufacturing
	334112	Computer Storage Device Manufacturing
33421		Telephone Apparatus Manufacturing
	334210	Telephone Apparatus Manufacturing
33422		Radio and Television Broadcasting and Wireless Communications Equipment Manufacturing
	334220	Radio and Television Broadcasting and Wireless Communications Equipment Manufacturing
33429		Other Communications Equipment Manufacturing
	334290	Other Communications Equipment Manufacturing
33431		Audio and Video Equipment Manufacturing
	334310	Audio and Video Equipment Manufacturing
33441		Semiconductor and Other Electronic Component Manufacturing
	334411	Electron Tube Manufacturing
	334412	Bare Printed Circuit Board Manufacturing
	334413	Semiconductor and Related Device Manufacturing
	334414	Electronic Capacitor Manufacturing
	334416	Electronic Coil, Transformer, and Other Inductor Manufacturing

	334419	Other Electronic Component Manufacturing
33451		Navigational, Measuring, Electromedical, and Control Instruments Manufacturing
	334511	Search, Detection, Navigation, Guidance, Aeronautical, and Nautical System and Instrument Manufacturing
	334512	Automatic Environmental Control Manufacturing for Residential, Commercial, and Appliance Use
	334513	Instruments and Related Products Manufacturing for Measuring, Displaying, and Controlling Industrial Process Variables
	334514	Totalizing Fluid Meter and Counting Device Manufacturing
	334515	Instrument Manufacturing for Measuring and Testing Electricity and Electrical Signals
	334516	Analytical Laboratory Instrument Manufacturing
	334518	Watch, Clock, and Part Manufacturing
	334519	Other Measuring and Controlling Device Manufacturing
33461		Manufacturing and Reproducing Magnetic and Optical Media
	334612	Prerecorded Compact Disc (except Software), Tape, and Record Reproducing
33511		Electric Lamp Bulb and Part Manufacturing
	335110	Electric Lamp Bulb and Part Manufacturing
	335129	Other Lighting Equipment Manufacturing
33521		Small Electrical Appliance Manufacturing
	335211	Electric Housewares and Household Fan Manufacturing
33522		Major Appliance Manufacturing
	335221	Household Cooking Appliance Manufacturing
	335222	Household Refrigerator and Home Freezer Manufacturing
	335224	Household Laundry Equipment Manufacturing
	335228	Other Major Household Appliance Manufacturing
33531		Electrical Equipment Manufacturing
	335311	Power, Distribution, and Specialty Transformer Manufacturing
	335312	Motor and Generator Manufacturing
	335313	Switchgear and Switchboard Apparatus Manufacturing
33591		Battery Manufacturing
	335911	Storage Battery Manufacturing

Table 3.1 (Continued)

Five-digit	Six-digit	Definition
33592		Communication and Energy Wire and Cable Manufacturing
	335921	Fiber Optic Cable Manufacturing
	335929	Other Communication and Energy Wire Manufacturing
33599		All Other Electrical Equipment and Component Manufacturing
	335991	Carbon and Graphite Product Manufacturing
	335999	All Other Miscellaneous Electrical Equipment and Component Manufacturing

Source: Author's creation based on revised data from United States Census Bureau (http://www.census.gov/eos/www/naics/).

with 334 (Computer and Electronic Product Manufacturing) and 335 (Electrical Equipment, Appliance, and Component Manufacturing) in the NAICS. However, these codes do not cover air conditioners, which are classified under code 333 (Machinery Manufacturing). Therefore, our dataset includes 33341 (Ventilation, Heating, Air-Conditioning, and Commercial Refrigeration Equipment Manufacturing) and the classification codes beginning with 334 and 335. The six-digit classification codes and descriptions are listed in Table 3.1 to show all the product types included with the five-digit classification codes.

The descriptive statistics for our dataset are provided in Table 3.2. The average amount of sales of foreign firms is approximately 2.7 times larger than that of indigenous firms. In addition, the average number of firms in each sector is 762, including both indigenous and foreign firms. There is a significant difference between the number of firms in the smallest and the largest sectors.

Table 3.3 presents the variables used in our study. TFP (*tfp*) and value added (*va*) are used as dependent variables for the growth of indigenous firms. TFP is calculated through estimation of a firm-level production function with value

Table 3.2 Descriptive statistics for the dataset in 2007

	Number of firms	Sales (USD)			
		Mean	Minimum	Maximum	SD
All firms	28,948	33,905	0	25,597,879	310,936
Indigenous firms	20,557	21,835	0	12,425,221	201,860
Foreign firms	8,391	59,315	0	25,597,879	462,071
	Number of firms	Number of firms by sector			
		Mean	Minimum	Maximum	SD
Sector size	28,948	762	96	3,255	704

Source: Author's creation based on revised data from the ORIANA provided by Bureau van Dijk.

Table 3.3 List of variables

Variables	Denotation
Total factor productivity	*tfp*
Value added	*va*
Fixed assets	*k*
Number of employees	*l*
Costs of goods sold	*c*
Sales	*s*
Ratio of foreign firms in fixed assets	*rk*
Age	*age*
Dummy of the central region	*cent*
Dummy of the western region	*west*

Source: Author's creation.

added as output, and fixed assets (k), number of employees (l), and costs of goods sold (c) as inputs. Value added is calculated by subtracting the costs of goods sold from sales, that is, $va = s - c$.

The ratio of foreign firms' fixed assets to total fixed assets (rk) is the independent variable; we examine the coefficient of which in the present chapter.[4] The ratio used for this variable is calculated by dividing the fixed assets of foreign firms by those of indigenous firms. This ratio denotes the impact of inward FDI on the accumulation of fixed assets by both indigenous and foreign firms. We expect the effects of inward FDI to be lagged, so we use the ratio of foreign firms' fixed assets to total fixed assets (*rk*) for the previous year. In addition, we use sales (*s*) and ages of firms (*age*) as control variables. We also add regional dummy variables for the central region (*cent*) and the western region (*west*), and the eastern region is treated as the baseline.[5]

3.2.2 ESTIMATION OF PRODUCTIVITY

In this subsection, the technology gaps between indigenous and foreign firms and the relationship between these gaps and the business experience of indigenous firms will be examined. We use TFP as a proxy of the technological levels of indigenous and foreign firms. The Cobb–Douglas production function to estimate TFP is as follows:

$$Y_{it} = A_{it} K_{it}^{\beta} L_{it}^{\gamma}, \tag{3.1}$$

where Y is the amount of production as the output, K is the capital stock, L is labor, and A represents the state of technology and TFP. β and γ stand for the input shares of K and L, respectively, and i and t stand for sector and time, respectively. Making Equation (3.1) a log-linear model, we obtain the following equation:

$$\ln Y_{it} = \ln A_{it} + \beta \ln K_{it} + \gamma L_{it}. \tag{3.2}$$

Therefore, according to Equation (3.2), the TFP of sector i at time t can have the residual form:

$$\ln A_{it} = \ln Y_{it} - \beta \ln K_{it} - \gamma L_{it}. \tag{3.3}$$

We use the following equation based on Equation (3.3) to derive TFP:

$$\ln Y_{it} = \alpha + \ln A_{it} + \beta \ln K_{it} + \gamma L_{it} + \varepsilon_{it}, \tag{3.4}$$

where α is the constant term and ε is the error term. In our firm-level production function of Equation (3.4), we use the variables of value added (va), fixed assets (k), and number of employees (l) as the output and inputs of Y, K, and L, respectively.

However, an endogeneity problem is likely to occur from the interdependent relationships between input and output in the production functions. Therefore, we use the method of Olley and Pakes (1996) to avoid this problem and estimate the production function nonparametrically. Using their method, we can filter out the influences of productivity of each firm on the level of inputs, so that we can calculate each firm's TFP without the problem. However, the method developed by Olley and Pakes requires information on investment to estimate production functions, though small and medium-sized firms do not necessarily invest every year. To solve this problem, Levinsohn and Petrin (2003) developed a method using information of intermediate inputs instead of investment. Therefore, we apply the Levinsohn and Petrin method to use the variable for costs of goods sold (c) instead of a variable for investment.

First, the TFP level and the growth rate by ownership are shown in Table 3.4. The growth rate of TFP is derived as the log difference of the TFP levels in two periods. It is evident that the average TFP level of indigenous firms is lower than that of foreign firms in China.[6] Although the average TFP growth rate of indigenous firms is a little bit higher than that of foreign firms, indigenous firms do not appear to catch up to the TFP level of foreign firms rapidly (Table 3.4), despite the fact that the Chinese government encouraged the adoption of new technologies during our observation period.

In the second step, we calculate the TFP level of indigenous firms for each sector (Table 3.5). The TFP levels of indigenous firms are lower than those of foreign firms in every sector indicating the technology gaps between indigenous and foreign firms. The fourth column shows the technology gaps in terms of TFP differences, which is the ratio of the TFP level of foreign firms to that of indigenous firms.

Table 3.4 TFP levels and growth by ownership

Ownership	Mean
TFP level	
Indigenous firms	3.8245
Foreign firms	5.0991
TFP growth	
Indigenous firms	0.1614
Foreign firms	0.1586

Source: Author's creation.

Table 3.5 Productivities of indigenous and foreign firms by sector

Sector	Indigenous		Foreign		Technology gap	Ratio of experiences
	Mean	SD	Mean	SD		
333412	3.9086	1.3134	5.5828	1.3605	1.4283	15.6
333415	3.6577	1.1550	4.6652	0.9958	1.2755	33.4
334111	4.6140	1.5621	5.6071	1.8887	1.2152	82.2
334112	4.0534	1.2295	5.4516	1.4380	1.3450	95.8
334210	4.2556	1.3078	5.7852	1.7758	1.3594	80.2
334220	3.9699	1.3361	4.6217	1.2697	1.1642	14.0
334290	4.3043	1.3203	5.2872	1.4614	1.2284	26.8
334310	3.6944	1.2922	5.1178	1.4582	1.3853	71.9
334411	3.7197	1.4063	5.3850	1.6554	1.4477	58.6
334412	3.6910	1.1651	5.2961	1.3546	1.4348	91.5
334413	3.7935	1.2586	5.0394	1.4099	1.3284	76.4
334414	3.6761	1.1226	4.6613	1.0891	1.2680	51.3
334416	3.8832	1.2074	5.0570	1.3459	1.3023	46.6
334419	4.2855	1.3713	5.1533	1.4195	1.2025	52.5
334511	4.6944	1.0946	5.5671	1.0181	1.1859	4.4
334512	3.8883	1.1019	4.9567	1.1006	1.2748	17.5
334513	3.9684	1.1818	5.3393	1.2721	1.3455	37.3
334514	4.1630	1.0389	5.2321	1.3379	1.2568	45.6
334515	3.8247	1.2116	5.1254	1.1554	1.3401	42.9
334516	3.8719	0.9727	4.9724	1.2674	1.2842	41.5
334518	3.4955	1.1466	4.4250	1.2261	1.2659	64.6
334519	3.9929	1.3457	5.1508	1.0542	1.2900	19.6
334612	3.3228	1.2882	4.7021	1.6379	1.4151	49.4
335110	3.4319	1.0418	4.6455	1.0869	1.3536	46.7
335129	3.4494	0.9885	4.8391	1.1446	1.4029	54.3
335211	3.8459	1.1479	5.2373	1.2291	1.3618	40.1
335221	3.7985	1.0494	5.0999	1.4045	1.3426	70.0
335222	4.0618	1.5287	5.5063	1.7512	1.3556	26.9

335224	3.8820	1.2229	5.7335	1.6223	1.4769	72.2
335228	3.8416	1.2690	5.0304	1.3412	1.3094	29.4
335311	4.0249	1.2156	5.3531	1.4853	1.3300	27.6
335312	3.6590	1.2010	4.9854	1.3377	1.3625	26.4
335313	3.7967	1.1567	5.3003	1.4163	1.3960	24.0
335911	3.6203	1.2910	5.0350	1.4211	1.3908	53.5
335921	4.4427	1.2582	5.2848	1.2193	1.1895	39.6
335929	3.7262	1.2614	4.9509	1.2548	1.3287	27.6
335991	3.9055	1.2198	5.1068	1.2298	1.3076	12.4
335999	3.7349	1.1810	4.8981	1.2689	1.3114	32.6
Average	3.8245	1.2367	5.0991	1.3910	1.3333	44.8

Note: Ratio of experiences means sales of foreign firms to the total sales.
Source: Author's creation.

The average gap is 1.3333, indicating that the productivity of foreign firms is approximately 1.3 times higher than that of indigenous firms. Sector 335224 (Household Laundry Equipment Manufacturing) has the largest gap, 1.4769, and sector 334220 (Radio and Television Broadcasting and Wireless Communications Equipment Manufacturing) has the smallest gap, 1.1642. This exhibits the technology gaps within the electronics industry.

Finally, we examine the relationship between the technology gaps and the business experiences of indigenous firms and find a positive correlation. When we consider the gaps between first-movers of foreign firms in developed countries and latecomers of indigenous firms in developing economies, the concepts of the experience effect, the learning effect, and learning-by-doing (LBD) should be taken into account. The more a firm produces, the more it can decrease the average cost of production. However, we do not have sufficient long-term information on accumulated production volume in each sector for the early years in our dataset.[7] Therefore, we use a ratio of foreign firms in sales in each sector to show comparative experiences of indigenous firms. Viewing each sector as a product market, it is reasonable to expect that if indigenous firms can produce and sell more than foreign firms, then this ratio of experiences would decrease. The ratio certainly does not directly indicate the experiences of indigenous firms; however, it shows comparative experiences between foreign and indigenous firms. Similarly, if the production of indigenous firms relative to foreign firms increases, then the productivity gap would decrease. Although the relationship does not identify a direction of causality, it is reasonable to suppose that there is a relationship between the two.

We use the five-year (2003–2007) average value of the ratio of experiences to smooth fluctuation of the ratio of experiences (see the last column of Table 3.5). The average ratio of foreign firm sales to total sales is 44.8 percent. Therefore, over half of total sales are made by indigenous firms. Sector 334112 (Computer Storage Device Manufacturing) has the largest ratio, 95.8 percent, and the sector 334511 (Search, Detection, Navigation, Guidance, Aeronautical, and Nautical System and Instrument Manufacturing) has the smallest ratio, 4.4 percent (Table 3.5). This shows that there are significant differences in the relative experiences of indigenous and foreign firms.

The relationship between the technology gap and the foreign firm sales ratio by sector is shown in Figure 3.1. Although it, of course, does not exhibit any causal relationship, it has been found that the larger the ratio of foreign firms' sales to total sales, the greater the technology gaps in a given sector. That is, the smaller the relative production of indigenous firms, the greater the technology gaps. Based on these facts, in the next section, we will analyze the effects of inward FDI on the growth of indigenous firms.

3.3 ANALYZING THE EFFECTS OF INWARD FOREIGN DIRECT INVESTMENT

3.3.1 METHOD

We estimate the correlations between the growth of indigenous firms and the ratios of foreign firms' fixed assets to all fixed assets. Using a combination of dependent and

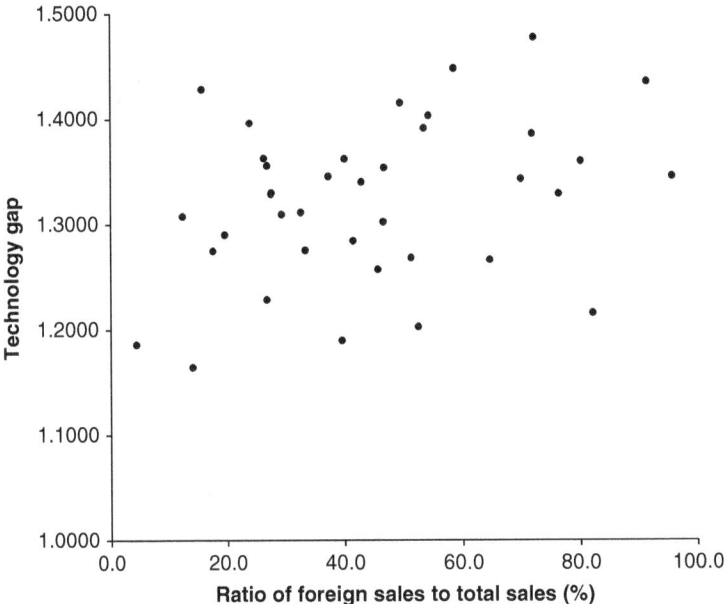

Figure 3.1 Relationship between the technology gaps and the ratios of foreign firms' sales to total sales
Source: Author's creation.

independent variables, we estimate the relationship between indigenous firms' growth and inward FDI as follows:

$$\ln va_{it} = \alpha + \beta rk_{it-1} + x'_{it}\gamma + \varepsilon_{it}, \qquad (3.5)$$

$$\ln va_{it} = \alpha + \beta rk_{it-1} \times industrialdummy + x'_{it}\gamma + \varepsilon_{it}, \qquad (3.6)$$

$$\ln tfp_{it} = \alpha + \beta rk_{it-1} + x'_{it}\gamma + \varepsilon_{it}, \qquad (3.7)$$

$$\ln tfp_{it} = \alpha + \beta rk_{it-1} \times industrialdummy + x'_{it}\gamma + \varepsilon_{it}, \qquad (3.8)$$

where i indicates firm and t indicates year. The two dependent variables are value added (va) in Equations (3.5) and (3.6), and TFP (tfp) in Equations (3.7) and (3.8). The two independent variables are the ratio of foreign firms' fixed assets to total assets (rk) in Equations (3.5) and (3.7), and a cross-term dummy variable (*industrialdummy*) for industry sector in Equations (3.6) and (3.8). In addition, we add control variables of sales (*s*) and ages of firms (*age*) and regional dummy variables (*cent* and *west*), as *x*.

We use random-effects generalized least squares (GLS) regression for our estimation. Because the ages and the regional dummy variables take constant values over time, we cannot use a method that assumes fixed effects. Moreover, we use GLS to avoid problems with heteroskedasticity in the error terms.

3.3.2 RESULTS

Estimation results for Equations (3.5)–(3.8) are shown in Table 3.6. In Equation (3.5), we estimate a regression of ln *va* on *rk*, with a one-year lag and control variables. The table shows that inward FDI has a negative effect on the electronics industry on the whole, at a significance level of 1 percent for the coefficient of value added. Consequently, the expansion of foreign firms, indicated by the fixed assets ratio, decreases the level of value added for the next year. Therefore, this result detects the market-stealing effect. The coefficients of ln *s* and ln *age* are positive at significant levels, indicating that larger and older indigenous firms tend to create more value added. The dummy variables are positive, indicating that indigenous firms in the central and western regions tend to create more value added in comparison to firms in the eastern region. The eastern region is considered to have a much higher number of small and medium-sized firms than the other regions, because the business environment in the east is well developed. Equation (3.5) indicates a negative effect of inward FDI on industrial development. Therefore, we examine the positive and negative effects at the sector level and investigate reasons for the different effects.

We estimate the effect of inward FDI on value added in each sector in Equation (3.6) to determine the different effect in the electronics industry. Table 3.6 shows that inward FDI has the significant effects in some sectors (e.g., 334112,

Table 3.6 Results of estimation

Equation	ln va		ln tfp	
	(3.5)	(3.6)	(3.7)	(3.8)
rk	−0.0047 (0.001)***		−0.0045 (0.001)***	
Cross term				
333412		−0.0084 (0.037)		−0.0099 (0.037)
333415		−0.0009 (0.007)		−0.0035 (0.007)
334111		−0.0007 (0.003)		−0.0031 (0.003)
334112		−0.0102 (0.002)***		−0.0080 (0.002)***
334210		−0.0077 (0.004)*		−0.0067 (0.004)*
334220		−0.0059 (0.022)		−0.0077 (0.021)
334290		−0.0067 (0.007)		−0.0066 (0.008)
334310		−0.0121 (0.004)***		−0.0100 (0.004)***
334411		−0.0024 (0.008)		−0.0075 (0.011)
334412		−0.0064 (0.002)***		−0.0057 (0.002)***
334413		0.0020 (0.002)		0.0008 (0.002)
334414		0.0102 (0.007)		0.0106 (0.007)
334416		−0.0026 (0.006)		−0.0014 (0.007)
334419		0.0012 (0.003)		−0.0026 (0.003)
334511		0.3496 (0.380)		0.4814 (0.376)
334512		0.0028 (0.020)		−0.0104 (0.024)
334513		−0.0076 (0.010)		0.0068 (0.010)
334514		0.0074 (0.016)		0.0008 (0.015)
334515		0.0244 (0.013)*		0.0083 (0.014)
334516		−0.0052 (0.019)		0.0261 (0.022)
334518		0.0126 (0.012)		0.0088 (0.012)
334519		0.0007 (0.022)		−0.0011 (0.021)
334612		−0.0175 (0.010)*		−0.0267 (0.010)**
335110		−0.0151 (0.006)**		−0.0104 (0.006)
335129		0.0091 (0.010)		0.0031 (0.010)
335211		−0.0109 (0.008)		−0.0223 (0.010)**
335221		−0.0094 (0.005)*		−0.0091 (0.005)*
335222		−0.0221 (0.017)		−0.0257 (0.017)
335224		−0.0075 (0.006)		−0.0033 (0.006)
335228		−0.0078 (0.006)		−0.0089 (0.006)
335311		0.0119 (0.015)		0.0086 (0.014)
335312		−0.0072 (0.013)		−0.0147 (0.013)
335313		0.0033 (0.012)		−0.0018 (0.013)
335911		−0.0148 (0.004)***		−0.0131 (0.004)***
335921		−0.0047 (0.007)		0.0016 (0.007)
335929		−0.0152 (0.007)**		−0.0076 (0.007)
335991		−0.0116 (0.028)		−0.0354 (0.027)
335999		−0.0045 (0.006)		−0.0065 (0.006)
lns	0.9228 (0.013)***	0.9124 (0.014)***	0.6556 (0.015)***	0.6529 (0.015)***
lnage	0.1731 (0.069)**	0.1414 (0.070)**	−0.0023 (0.068)	−0.0282 (0.071)
Dummy				
cent	0.2344 (0.103)**	0.2010 (0.107)*	0.1159 (0.101)	0.1315 (0.107)
west	0.2828 (0.110)***	0.2642 (0.112)**	0.0934 (0.108)	0.1171 (0.113)
Constant	−1.1610 (0.214)***	−0.9791 (0.225)***	−1.1866 (0.224)***	−1.0834 (0.245)***
Sample size	20557	20557	20557	20557
R^2	0.8700	0.8862	0.7719	0.7928

Notes: Standard errors are in parentheses. ***, ** and * represent statistical significance at the 1, 5 and 10 percent, respectively.
Source: Author's creation.

334210 and 334310). These effects are almost entirely negative, except for sector 334515 (Instrument Manufacturing for Measuring and Testing Electricity and Electrical Signals). Some other sectors show the positive effects, but not at a significant level. All the other independent variables reveal similar results in comparison with Equation (3.5). Consequently, in some sectors, inward FDI has a negative effect on the value added of indigenous firms.

We estimate the effect of inward FDI on the TFP level in Equations (3.7) and (3.8). As in Equation (3.5), the results of Equation (3.7) show that inward FDI has a negative effect on the TFP level of indigenous firms. The control variables are not significant, except that sales (s) are significant. This shows that there is a negative effect of inward FDI on the TFP level of indigenous firms in the electronics industry.

When we estimate the effect of inward FDI on the TFP level in each sector in Equation (3.8), we find that, in some sectors, inward FDI has significant effects. Moreover, all of the significant sectors have negative effects. These facts also indicate that inward FDI has negative effects on the TFP levels of indigenous firms in some sectors.

As the above results show, inward FDI has a negative effect on the value added and the TFP level for the electronics industry as a whole. However, by breaking down the industry, we find that there are different effects, mostly negative or zero, for the 38 sectors. Therefore, we will investigate factors contributing to the difference in the next subsection. We could not find any positive effects of inward FDI at significant levels in many sectors. However, this does not mean that no technology spillover effect is occurring in the industry. The positive effect might be found when bigger and different datasets are used.

3.3.3 Discussion

In this subsection, we discuss the relationships between the sectors' technology gaps and the ratios of experiences discussed. In Section 3.2, we have shown the technology gap and the ratio of foreign firms' sales to the total sales for each sector. In this subsection, we will discuss relationships between these characteristics and the positive and negative effects found through our regression models. Table 3.7 shows the relationship between these characteristics and the effects of inward FDI, sorting by the size of the technology gaps in Table 3.7(a) and by ratios of foreign firms' sales to the total sales in Table 3.7(b), respectively.

As shown in Table 3.7(a), inward FDI tends to have a negative effect on sectors with larger technology gaps. Indigenous firms that are technologically lagging cannot absorb technologies effectively, because technologies brought by foreign firms may be too advanced for indigenous firms. However, at the same time, the technological lags provide more room for technological progress for indigenous firms.

This same trend also holds in Table 3.7(b), which is sorted by foreign firms' sales ratio. The results show that indigenous firms lacking in business experience tend to experience a negative effect from the expansion of inward FDI. In other words, too large a difference in business experience results in inhibition of growth for indigenous firms.

In fact, indigenous firms are likely to suffer the negative effects of inward FDI when foreign firms have higher sales shares in comparison with indigenous firms. Table 3.8 shows the relationship between the sales shares of the top ten firms in each sector and the inward FDI effects. The first to fifth columns are the same as those in Table 3.7(a). The sixth column is the concentration ratio of the top ten firms in sales (*CR10*) for each sector and the last column is the share of foreign firms in the *CR10*. The concentration ratio is the ratio of total sales of the top ten firms in all of sales in a sector. As shown in Table 3.8, we do not find any obvious relationship between *CR10* and inward FDI effects; however, sectors with higher shares of foreign firms in the top ten tend to experience negative effects. Consequently, when indigenous firms face the lack of technology and the lack of experience, then they have a possibility to suffer a negative effect.

Table 3.7 Relationship between sector characteristics and inward FDI effects

(a) Technology gaps and inward FDI effects

(b) Ratios of experiences and inward FDI effects

Sector	ln *va*	ln *tfp*	Technology gap	Ratio of experiences	Sector	ln *va*	ln *tfp*	Technology gap	Ratio of experiences
335224			1.4769	72.2	334112	– – –	– – –	1.3450	95.8
334411			1.4477	58.6	334412	– – –	– – –	1.4348	91.5
334412	– – –	– – –	1.4348	91.5	334111			1.2152	82.2
333412			1.4283	15.6	334210	–	–	1.3594	80.2
334612	–	– –	1.4151	49.4	334413			1.3284	76.4
335129			1.4029	54.3	335224			1.4769	72.2
335313			1.3960	24.0	334310	– – –	– – –	1.3853	71.9
335911	– – –	– – –	1.3908	53.5	335221	–	–	1.3426	70.0
334310	– – –	– – –	1.3853	71.9	334518			1.2659	64.6
335312			1.3625	26.4	334411			1.4477	58.6
335211		– –	1.3618	40.1	335129			1.4029	54.3
334210	–	–	1.3594	80.2	335911	– – –	– – –	1.3908	53.5
335222			1.3556	26.9	334419			1.2025	52.5
335110	– –		1.3536	46.7	334414			1.2680	51.3
334513			1.3455	37.3	334612	–	– –	1.4151	49.4
334112	– – –	– – –	1.3450	95.8	335110	– –		1.3536	46.7
335221	–	–	1.3426	70.0	334416			1.3023	46.6
334515	+		1.3401	42.9	334514			1.2568	45.6
335311			1.3300	27.6	334515	+		1.3401	42.9
335929	– –		1.3287	27.6	334516			1.2842	41.5
334413			1.3284	76.4	335211		– –	1.3618	40.1
335999			1.3114	32.6	335921			1.1895	39.6
335228			1.3094	29.4	334513			1.3455	37.3
335991			1.3076	12.4	333415			1.2755	33.4
334416			1.3023	46.6	335999			1.3114	32.6
334519			1.2900	19.6	335228			1.3094	29.4
334516			1.2842	41.5	335311			1.3300	27.6
333415			1.2755	33.4	335929	– –		1.3287	27.6
334512			1.2748	17.5	335222			1.3556	26.9
334414			1.2680	51.3	334290			1.2284	26.8
334518			1.2659	64.6	335312			1.3625	26.4
334514			1.2568	45.6	335313			1.3960	24.0
334290			1.2284	26.8	334519			1.2900	19.6
334111			1.2152	82.2	334512			1.2748	17.5
334419			1.2025	52.5	333412			1.4283	15.6
335921			1.1895	39.6	334220			1.1642	14.0
334511			1.1859	4.4	335991			1.3076	12.4
334220			1.1642	14.0	334511			1.1859	4.4

Notes: The second and third columns in Tables 3.7 (a) and (b) show the positive and negative effects and the significant levels from Table 3.6. – – – (+ + +), – – (++), and – (+) show negative (positive) effects of inward FDI at significant levels of 1, 5, and 10 percent, respectively. Empty spaces mean that there is no significance, regardless of whether the effects are positive or negative.
Source: Author's creation.

Table 3.8 Sales shares in the top ten firms and inward FDI effects

Sector	ln va	ln tfp	Technology gap	Ratio of experiences	CR10	Share of foreign firms in the top 10
335224			1.4769	72.2	48.3	100.0
334112	− − −	− − −	1.3450	95.8	45.7	100.0
334111			1.2152	82.2	74.9	95.4
334412	− − −	− − −	1.4348	91.5	40.8	91.7
334413			1.3284	76.4	32.0	90.2
335221	−	−	1.3426	70.0	46.8	86.5
334612	−	− −	1.4151	49.4	51.4	74.0
334518			1.2659	64.6	31.2	73.6
334210	−	−	1.3594	80.2	73.6	72.3
334515	+		1.3401	42.9	31.3	71.4
335911	− − −	− − −	1.3908	53.5	31.7	65.2
334310	− − −	− − −	1.3853	71.9	47.1	62.2
335129			1.4029	54.3	41.8	59.2
334414			1.2680	51.3	55.6	58.8
334516			1.2842	41.5	41.8	57.2
334514			1.2568	45.6	62.9	55.3
335110	− −		1.3536	46.7	19.2	53.2
335929	− −		1.3287	27.6	12.6	50.9
334411			1.4477	58.6	83.4	44.5
334416			1.3023	46.6	25.0	38.7
334513			1.3455	37.3	28.6	38.6
334419			1.2025	52.5	34.8	37.6
335311			1.3300	27.6	46.7	34.6
335211		− −	1.3618	40.1	51.9	30.7
335921			1.1895	39.6	59.1	29.4
334290			1.2284	26.8	74.3	25.9
335222			1.3556	26.9	83.1	24.1
335999			1.3114	32.6	41.7	22.0
335312			1.3625	26.4	32.7	18.2
334519			1.2900	19.6	69.9	15.9
333415			1.2755	33.4	55.5	15.6
335228			1.3094	29.4	56.5	15.3
333412			1.4283	15.6	50.3	15.3
334512			1.2748	17.5	26.7	13.1
335313			1.3960	24.0	36.7	9.3
334220			1.1642	14.0	76.8	3.1
335991			1.3076	12.4	37.3	0.0
334511			1.1859	4.4	55.6	0.0

Notes: CR10 means the top ten firms in sales. − − − (+ + +), − − (++), and − (+) show negative (positive) effects of inward FDI at significant levels of 1, 5, and 10 percent, respectively. Empty spaces mean that there is no significance, regardless of whether the effects are positive or negative.
Source: Author's creation.

3.4 CONCLUSION

In this chapter, we have investigated the effect of inward FDI on the growth of indigenous firms. Our results show that inward FDI is likely to have a negative effect on the growth of indigenous firms in sectors with large gaps in technology and less experience in business. The main points of our study can be summarized as below.

First, in Section 3.2, we discussed how the characteristics of different sectors affect productivity. We found that every sector in the electronics industry faces a technology gap, which varies in size. Therefore, we compare the technology gap with the ratio of experiences by sector. Although this relationship requires a more rigorous analysis, it seems reasonable to say that the business experience of indigenous firms correlates to the technology gap in the sector on the whole.

Next, we conducted regression analyses to determine the effect of inward FDI on the growth of indigenous firms. Our analyses have shown that inward FDI has different effects on the growth of indigenous firms in terms of value added and TFP; however, we found negative effects of inward FDI in some sectors.

Finally, we compared our findings on the technology gaps and foreign firms' sales with the findings from the regression model. We have found that there is a relationship between the negative effects of inward FDI and the technology gaps. Sectors with large technology gaps tend to suffer greater negative effects. Moreover, based on the relationship between the technology gap and business experience, it is possible to say that sectors with lower levels of experience tend to be negatively impacted by inward FDI.

Our results show that China's electronics industry has been developing remarkably, but young indigenous firms might suffer a negative effect from the expansion of inward FDI. At the same time, the indigenous firms have plenty of capacity for future growth. Therefore, if indigenous firms are not able to find markets with no competition from foreign firms or to find strategies to compensate for the technology gap, young indigenous firms will not be likely able to successfully enter markets and achieve growth. In the next chapter, we will investigate the growth process of Chinese electronics firms facing the technology gap.

CHAPTER 4

DIVERSIFICATION MECHANISM

4.1 INTRODUCTION

How do indigenous firms in developing countries achieve growth when they confront a technology gap between themselves and foreign firms from developed countries?[1] We attempt to answer the question by examining the behavior of China's mobile phone handset firms, which have managed to compensate for the technology gap by differentiating their organization from foreign firms. To do this, we investigate the boundaries of indigenous firms under the influence of competitors of foreign firms.

Major electronics manufacturers have been achieving growth by using external technology and internal knowledge. A major home appliance manufacturer, Haier Group (Haier), has realized growth by buying key components from outside indigenous and foreign firms, and by providing meticulous after-the-sale services all over the country. They have set up nationwide sales and after-the-sale service networks down to the prefecture level, although they are also using chain retail stores especially in the urban markets. A major personal computer (PC) manufacturer, Lenovo, has been growing by focusing on each stage in the order of sales, manufacturing, and technology (*mao-gong-ji*) (Ling, 2005). Lenovo's founder, Mr. Chuanzhi Liu, initially focused on selling PCs and, using the seed money accumulated from the business, began developing the technology. Moreover, a major telecommunication equipment manufacturer, Huawei Technologies (Huawei), has been growing rapidly by selling price-competitive products in comparison with those of the Western manufacturers and by providing detailed after-the-sale services.[2] Indigenous firms have realized growth by capitalizing on business opportunities in the domestic market. And some of the major indigenous firms have now become global players.

To investigate the growth of Chinese firms, we take China's handset industry for the period of the 1990s up to 2008 as our case study. Indigenous handset firms attained rapid growth and expanded their market in tough competition with foreign firms in this period. Although the production of handsets certainly required advanced technological capabilities because at that time handsets were new products in the market, indigenous handset firms have grown successfully, especially since the late 1990s. This case study will illustrate the growth of indigenous handset firms in the embryonic period of industrial development because evaluating growth in the midst of changes can be difficult. Therefore, we do not include in this study recent changes after the full-scale introduction of the third generation of mobile phone standard (3G) and the smartphone boom. Therefore, this case of China's handset industry can be appropriate to our study, since it is difficult for us to evaluate the growth of indigenous firms in industries in the midst of changes. In China, the Global System for Mobile Communications (GSM) and Code Division Multiple Access (CDMA) as the second generation of mobile phone standard (2G), and Time Division–Synchronous Code Division Multiple Access (TD-SCDMA) as 3G, were in use as of 2008. We have chosen the GSM system as our case because it had the largest number of subscribers among the systems.[3] If not otherwise specified, handsets discussed in this chapter are for the GSM system.

In particular, this study focuses on the behavior of Chinese handset firms that they depended on outside firms and efficiently sold inexpensive handsets in local and rural markets, despite a significant technology gap. Although handsets had been partially modularized, it has been still difficult for Chinese firms to have a good command of key components of handsets for new model development. Therefore, they themselves decided to buy handsets and design services from outside firms. At the same time, in the fast-growing handset market, indigenous firms found a great opportunity to sell simple entry-level models to new subscribers in local and rural markets. They had swiftly organized nationwide sales networks in the markets. In the case of the handset business, it was effective for rapid growth to find out consumers' needs and the potential in local and rural markets through labor-intensive marketing. Since these features must be shared in other emerging countries, the case of the handset business can be a lesson for their indigenous firms.

In this chapter we conduct a qualitative case study. In addition to a literature review, eight field researches on this topic were conducted to comprehend the emerging industry in China.[4] The interviews had covered indigenous and foreign electronics manufacturers, chip vendors, software companies, independent design houses (IDHs) for handsets, distributors, retailers, marketing and consulting firms, telecommunication carriers, related specialists, and so on, in order to comprehend the emerging industry. In this chapter, we especially focus on the behavior to balance the three factors. As shown in Chapter 1, we have found that many previous studies revealed the facts that Chinese firms have attained rapid growth through using external technology and internal knowledge. Based on the previous studies, we are investigating the rationality of the behavior of Chinese firms and the optimal balance between the three factors, in order to develop the diversification mechanism.

The remainder of the chapter is organized as follows. In the next section, we review the growth pattern of indigenous firms and provides an introduction to the development of China's handset industry. Sections 4.3 and 4.4 present an investigation into the relationship between the technology gap and the boundaries of indigenous firms in two industrial development phases. The final section concludes our analysis.

4.2 Growth and Organizations

In this section we review the relationship between the growth process of indigenous firms and their boundaries. The growth process can be divided into three phases on the basis of changes in the total market share of indigenous firms. The boundaries of indigenous firms have also been corresponded to the changes in the market share. In this chapter we use the market share as the indicator of the growth of indigenous firms, although output, value added, profits, and so on, are mostly used as the measure of the growth. We have chosen this indicator since we are interested primarily in the growth of indigenous firms in comparison with that of foreign firms as competitors. Thus, using the market share, this chapter attempts to identify the indigenous and foreign firms are growing comparison to each other.

4.2.1 First Phase

The market was almost entirely dominated by foreign firms during the first phase, which lasted until 1998, though the market share of Chinese handset firms has increased since 1999.[5] About 80 percent of the market was held by three major foreign firms: Motorola (the United States (US)), Nokia (Finland), and former Ericsson (Sweden).[6] The remaining market share was accounted for by foreign firms such as Siemens (Germany), Philips (the Netherlands), NEC (Japan), former Matsushita Electric Industrial (Japan), and so on.[7]

Although the Chinese government and certain indigenous firms sought to domestically manufacture handsets, the domestic production did not succeed commercially. Some Chinese firms had assembled handsets through contract manufacturing with foreign firms. In addition, the Chinese government had facilitated a nationalization project that involved selected indigenous manufacturers. However, at the time, there were significant gaps in the technological levels and the amount of capital between indigenous and foreign firms. In particular, indigenous firms did not have sufficient technological capabilities to develop and manufacture the key components of handsets, which resulted in a sharp increase in production costs (Economic Structure Reform and Economic Operation Office of Ministry of Information Industry of China, 2003). The end result was that the persistent technology gap prevented the indigenous firms from expanding their market share.

4.2.2 SECOND PHASE

In the second phase from 1999 to 2003, indigenous firms continued to expand their market share in terms of the number of handsets, from 5.3 percent in 1999 to 52.9 percent in 2003 (Figure 4.1).[8] This expansion went hand-in-hand with a rapid increase in the number of subscribers. There has been a sustained increase in the number of subscribers ever since the start of the mobile phone service in China, and the number of subscribers has grown swiftly, especially since the late 1990s (Figure 4.2). The number of subscribers had annually increased from 50–100 million people since 2001, and the total national number of subscribers numbered at more than 600 million in 2008.

What triggered the expansion of the market share of indigenous firms was the adoption of an industrial policy in 1999 that favored indigenous firms.[9] Concerned that indigenous firms might fail to seize business opportunities in the fast-growing market, the Chinese government introduced a license system as an entry barrier, provided subsidies to indigenous firms for research and development (R&D) expenditure, and enacted local manufacturing content requirements to foreign firms. Among those, the license system had a particularly

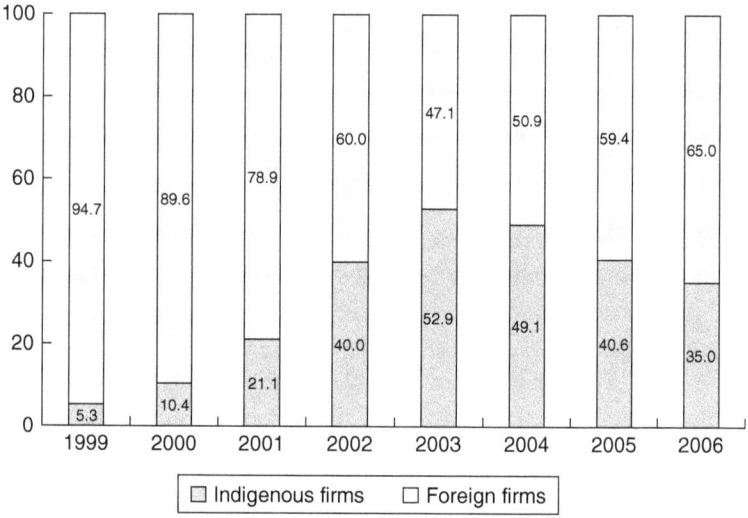

Figure 4.1   Market shares of indigenous and foreign firms, 1999–2006 (%)
Note: The share in 2005 is for January to September.
Source: Author's creation based on revised data from Ministry of Information Industry of China (2003) for 1999–2002, CCID (China Center for Information Industry Development) for 2003–2004, Ministry of Information Industry (2006) "2005 nian Woguo Shouji Chanye Fazhan Pingshu [Commentary on Development of Our Country's Mobile-Phone Handsets Industry in 2005]," accessed at http://www.mii.gov.cn/ on April 5, 2006 (in Chinese) for 2005, and Ministry of Information Industry of China (2007) for 2006.

significant benefit on indigenous firms to enter the market in the early development phase by blocking the new entry of foreign firms.[10] Later, the licensing was eased in 2005 and finally abandoned in 2007.

This protectionist policy led to an upsurge in the number of entries of indigenous firms into the industry. The great majority of the new entrants, however, did not have the technological capabilities and the experiences necessary to succeed in the handset business. Most of the new entrants came from the home appliance and consumer electronics industries, although telecommunication equipment manufacturers, such as Huawei Technologies (Huawei) and ZTE, were well equipped with the technological capabilities. In this chapter, we focus on major firms from the home appliance

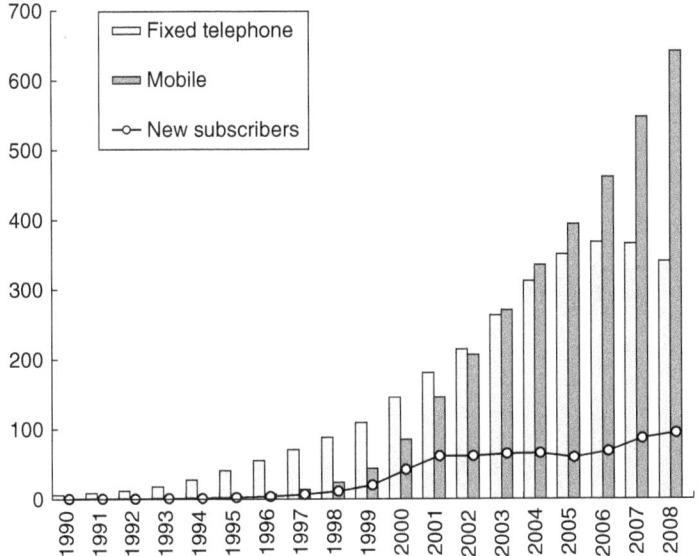

Figure 4.2 The number of subscribers, 1990–2008 (Million)
Note: Fixed telephone includes the Chinese Personal Handyphone System (PHS) system.
Source: Author's creation based on revised data from National Bureau of Statistics of China (various years).

and consumer electronics industries, such as Ningbo Bird (Bird), TCL, Konka Group (Konka), and Lenovo, which became handset manufacturers during the early years of the regulation. In addition, indigenous firms of various origin, such as distributors, also successively entered the industry one after the other. Moreover, the amount of illegal handsets in the market increased along with the rapid growth of the market.[11]

As a result, during the second phase, indigenous firms' market share determinably increased (Table 4.1). In this phase, indigenous firms primarily focused their business strategy on the sales stage (Figure 4.3). Figures 4.3 (a) and (b) illustrate the boundaries of indigenous and foreign firms, respectively. The black shaded sections represents the degree of areas internalized by firms on the value chain of handsets. Indigenous firms tended to focus on the sales stage than

Table 4.1 Market share by major firm, 1999–2008 (%)

	1999	2000	2001	2002	2003	2004	2005	2006	2007	2008
Indigenous firms										
Bird	n.a.	3.2	6.4	9.9	14.2	10.2	9.7	7.8	6.5	4.7
TCL	n.a.	1.0	3.0	8.7	11.2	6.5	2.7	2.2	1.7	1.8
Konka	n.a.	n.a.	n.a.	n.a.	6.2	5.8	0.6	0.4	0.3	0.5
Huawei	n.a.	n.a.	n.a.	n.a.	n.a.	n.a.	2.4	3.0	3.1	4.5
ZTE	n.a.	n.a.	n.a.	n.a.	n.a.	n.a.	2.2	2.5	3.1	4.3
Lenovo	n.a.	n.a.	n.a.	n.a.	n.a.	n.a.	1.7	1.7	1.9	2.7
Amoi	n.a.	n.a.	n.a.	n.a.	n.a.	n.a.	0.4	0.4	0.4	0.5
Foreign firms										
Nokia	32.3	25.1	22.3	18.2	11.1	15.0	33.4	35.6	35.7	34.8
Motorola	39.4	35.4	29.3	28.5	9.3	8.9	15.4	12.7	11.6	7.2
Samsung	n.a.	n.a.	n.a.	n.a.	n.a.	8.3	13.0	13.3	15.4	18.3
Sony Ericsson*	6.4	9.2	6.5	2.1	1.1	2.9	4.1	4.7	5.3	4.8
Philips	n.a.	n.a.	n.a.	n.a.	n.a.	2.8	0.6	0.6	0.4	0.4
Siemens	6.0	8.1	9.7	4.7	2.5	1.4	n.a.	n.a.	n.a.	n.a.

Note: The share of Sony Ericsson before October, 2001, is the share of Ericsson.
Source: Author's creation based on revised data from Ministry of Information Industry of China (various years) for 1999–2004, and Passport provided by Euromonitor International for 2005–2008.

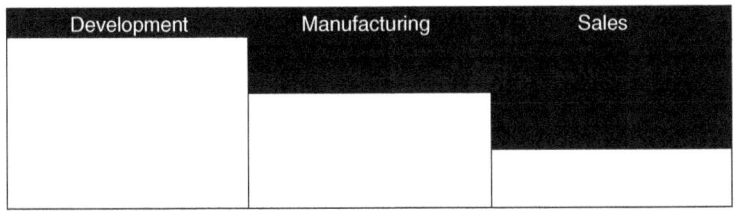

Figure 4.3 The boundaries of indigenous and foreign firms in the second phase (a) Indigenous firms (b) Foreign firms
Source: Author's creation based on various materials.

foreign firms. Lacking technological capabilities and experience, major indigenous firms bought many handsets from firms in South Korea (Korea) and Taiwan and sold them in the protected Chinese domestic market, although major Chinese handset firms also imported production lines and assembled handsets at the same time.[12] Foreign firms also bought some of the handsets from contract manufacturers to enrich their product lineups. In contrast to the sales stage, indigenous firms almost ignored focusing on the development stage, except designing appearances of handsets. On the other hand, foreign firms focused on developing new models in order to differentiate their products, although they also did not manufacture some of the key components.

4.2.3 Third Phase

The last phase began after 2004. With increased competition, the market share of indigenous firms dropped to about

40 percent. In this phase, although indigenous firms continued to depend on outside firms for developing and designing new models, they sought to integrate partially the product development stage.

Although the market share had dropped for some of the indigenous firms, Lenovo had increased its share. Although the lack of technology made it difficult for Lenovo to stay at the forefront of the market expansion, nevertheless the firm achieved growth by incorporating marketing capabilities developed through their PC business. Figures 4.4 (a) and (b) illustrate the boundaries of indigenous and foreign firms, respectively. In this phase, indigenous firms sought to internalize a larger part of the development stage to differentiate their products and successfully adapt to the increasing competition. Foreign firms internalized a greater part of the sales stage to absorb the competitiveness of indigenous firms, aiming to expand sales in local and rural markets.

(a)
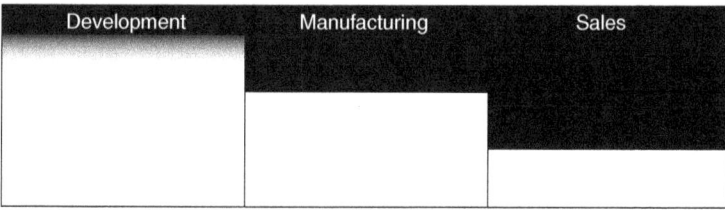

(b)
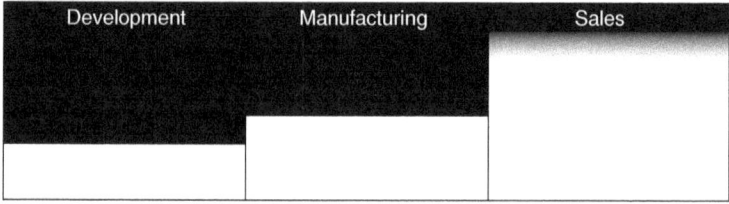

Figure 4.4 The boundaries of indigenous and foreign firms in the third phase (a) Indigenous firms (b) Foreign firms

Source: Author's creation based on various materials.

In the following sections, we analyze the make-or-buy decision of indigenous firms under the technology gap in the second and third phases. We skip the first phase, because the boundaries of indigenous firms at the phase do not almost exist.

4.3 ANALYSIS OF THE SECOND PHASE

4.3.1 PRODUCT STRUCTURE AND TECHNOLOGY GAP

In China's handset industry, there was still a technology gap between indigenous and foreign firms, although technologies related to handsets had somewhat matured and the product structure had been partially simplified. The GSM system that is prevalent in China has been used mainly in European countries since the mid-1990s. Therefore, the technologies had partially matured and some key components had been modularized. However, the product structure was still complicated in comparison with other consumer electronics such as desktop PCs. In other words, the technology gap between indigenous and foreign firms was closely related to the complexity of the composition of the product.

In terms of product structure, a handset is composed of hardware and software, each of which can be divided into three layers (Figure 4.5). Hardware comprises the following layers: (1) the core layer, which mainly contains a radio frequency (RF) device for communication functions and information processing for signals; (2) the middle layer, which includes various devices mounted on a printed circuit board (PCB); and (3) the surface layer, which comprises an outer case of handsets and a keypad. Meanwhile software comprises (1) the core layer, which mainly contains an operating system (OS) for the basic software; (2) the middle layer, consisting of middleware for communication functions; and (3) the surface layer, which includes a user interface and a variety of application software.

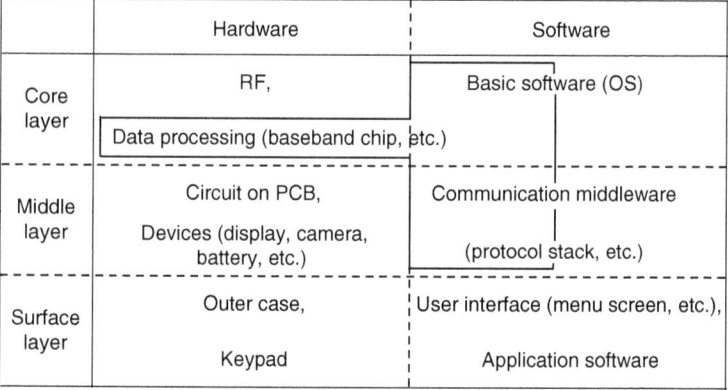

Figure 4.5   Product structure
Note: The elements contained within the bold line are modularized as platforms.
Source: Author's creation based on various materials.

As we can see from Figure 4.5, the structure is complicated, but its intricacy has been simplified to a certain extent by partial modularization. The information processing element and the basic components of the OS and the middleware, all of which are contained within the bold line in the figure, have been modularized as platforms developed by major chip vendors in the US (such as Texas Instruments) and the European Union (EU).[13] Major foreign firms also use platforms for new model development and design.

The platforms had given new entrants an advantage over advanced technologies needed to independently develop and design the platforms. However, they still required sufficient development and design experiences to have a good command of platforms to differentiate products in an increasing range of handset functions. Moreover, platforms developed by chip vendors in developed countries required advanced technological capabilities, but sufficient technical support was not provided to inexperienced newcomers (interview at Konka in Shenzhen, on July 26, 2006). Thus, even though ready-made platforms were available, platforms could not completely cover the limited technological capabilities of indigenous firms.

The lack of technology of indigenous firms was reflected in the situation of China's handset industry in the 1990s. As mentioned above, the Chinese government conducted the nationalization project for the domestic handset industry, but the initiative did not achieve commercial success. Moreover, some major Chinese electronics manufacturers attempted to enter the industry and to expand their market share, but gained only a foothold, and were unable to achieve large-scale production. Therefore, indigenous firms did not have sufficient technological capabilities when they entered the market.

4.3.2 EXTERNAL TECHNOLOGY

As indigenous firms lacked technological capabilities, they turned to outside firms for product development and, in some cases, manufacturing, and bought handsets mainly from original equipment manufacturers (OEMs), original design manufacturers (ODMs), and IDHs in Korea and Taiwan. IDHs are firms that provide services of product design and development. Thus, depending on outside firms, indigenous firms increased product lineups of inexpensive handsets in comparison with foreign firms' handsets. For example, at the time, Bird used to buy handsets from Pantech (Korea), Sewon Telecom (Korea), BenQ (Taiwan), Quanta Computer (Quanta) (Taiwan), and so on, whereas TCL used to purchase from Pantech (Korea), LG Electronics (Korea), Hon Hai Precision Industry (Hon Hai) (Taiwan), and so on. Apparently, in 2003, almost two-thirds of the handsets marketed by indigenous firms came from Taiwan. Korean and Taiwanese firms provided almost finished products, and Chinese firms, protected by the government licensing system, sold them under their own brand names.

While Chinese firms heavily depended on outside firms, the make-or-buy decision had at least two economic rationalities. First, although Chinese firms had to start accumulating technologies and know-how for the information and communication technology (ICT) business, but Korean and Taiwanese firms already possessed them through transactions with foreign firms. For example, Hon Hai and Quanta established in 1974 and 1988, respectively, had accepted many orders from Dell (the US), Hewlett-Packard (the US), Sony (Japan), and so on, and had built capabilities for ICT-product development and manufacturing. Transactions with firms in developed countries had given Taiwanese firms opportunities for learning and building technological capabilities (Kawakami, 2012; Kawakami and Sturgeon, 2011). In addition, Korean firms also had built capabilities through transactions with firms in developed countries. For example, Pantech, established in 1991, accepted numerous orders from Motorola and gained know-how in the handset business (Abe, 2006). Consequently, indigenous Chinese firms could run their handset businesses by depending on external technology, because indigenous firms did not have enough capabilities for product development and manufacturing. As a result, Korean and Taiwanese firms also could increase sales in China through increased transactions with indigenous Chinese firms, while their direct entries in the Chinese market were blocked by the protectionist industrial policy.

Second, simple and low-cost handsets were in demand in the rising stage of the Chinese market. Therefore, Chinese firms had no need to offer well-differentiated and sophisticated handsets to new subscribers, as simple basic specifications were enough to meet the market demand at that time. The basic specifications comprised telephone call and short message service (SMS) functions. Just simple and inexpensive handsets incorporating these two functions were needed for new users. Consequently, indigenous firms could depend on external technology, although outside firms would not make efforts to customize handsets only for specific Chinese firms, which could not expect business sustainability of future sales.

4.3.3 Internal Knowledge

Although indigenous Chinese firms depended on Korean and Taiwanese suppliers, it was not as if they did nothing at that time. They pursued the marketing-oriented strategy, using the inherent home advantage, which worked in their favor as long as they operated in the home market. Indigenous firms recognized local and rural markets as business opportunities to sell inexpensive handsets.[14] In particular, some major indigenous handset firms, such as Bird, TCL, Konka, and so on, optimized this strategy (Lu et al., 2006). Although they did not have product development capabilities, they designed only handsets' surface layers, which could satisfy Chinese consumers' tastes. Moreover, a distinguishing strategic feature of indigenous firms was that they independently organized their own sales networks.

There was a significant difference between the distribution policies of indigenous and foreign firms. Figure 4.6 shows the general distribution channels for GSM handsets in China, and columns (1) and (2) show the distribution channels followed by foreign and indigenous firms, respectively.[15] A significant difference is seen between the distribution stages of foreign and indigenous firms. Handsets produced by foreign firms entered the market through nation-level distributors to the province- and prefecture-level distributors and retailers, which were included by the independent decision of distributors at each stage.[16] Thus, after shipping the products to a small number of nation-level distributors, foreign firms were excluded from distribution. With this distribution policy, firms were relieved from exercising cost control over the distribution channels; however, they were unable to control the distribution margins, which tended to increase along with each distribution stage.

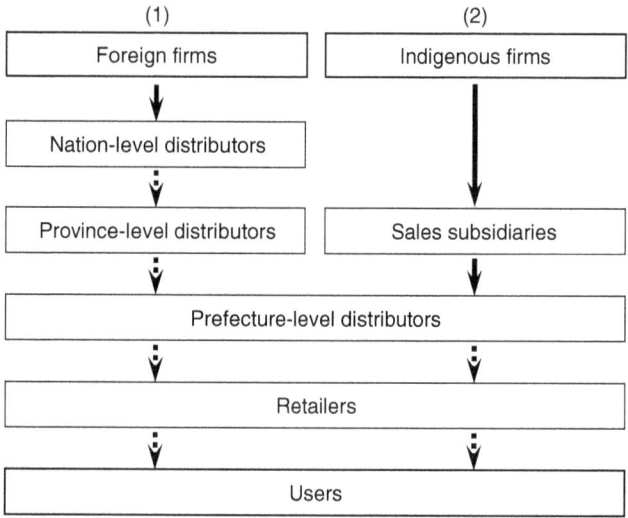

Figure 4.6 Distribution channels

Note: The dashed arrows indicate transactions outside firms' boundaries; the solid arrows indicate transactions inside firms' boundaries. However, in the case of indigenous firms, they engaged in every distribution stage.
Source: Author's creation based on various materials.

In contrast, major Chinese firms were closely involved with the distribution channels. They established sales subsidiaries at the province level, and these subsidiaries executed various sales policies, although the details differed by firm. Generally, subsidiaries selected prefecture-level distributors and monitored their behavior of these distributors, especially in matters such as pricing and choosing subsequent distributors. They controlled distribution channels in order to prevent price reductions due to competition among distributors that overlapped other business spheres. Moreover, subsidiaries dipatched sales promoters to retail shops to expand the sales of their own brand handsets. Indigenous firms promptly conducted the labor-intensive marketing. Bird in particular was strongly committed to the marketing-oriented strategy and substantially expanded its market share during this phase (interview at Bird in Ningbo, on August 31, 2004). For example, in 2000 they set up 28 subsidiaries in provincial capitals and 300 offices in local major cities to cover all of the country.

Indigenous firms had the rationality to organize own sales networks from the viewpoint of the boundaries of the firm. To increase sales capabilities, firms needed to collect and analyze market information on each distribution stage, and to train salespersons and promoters. Foreign firms had already established close relationships with major nation-level distributors since the mid-1990s, and this made it difficult for indigenous firms to sell their products due to the existing major distributors (Lin Huang, 2003). Meanwhile, the distributors were wary of devoting their human resources to transactions with specific indigenous firms because of the uncertainty of sales of local brand products. Thus, indigenous firms tended to be trapped in a vicious circle in which their products were not distributed because of the anticipation of poor sales. Therefore, major indigenous firms, such as Bird, TCL, and so on, organized their own sales networks to mobilize human resources for the expansion of sales.

Although the rationale of partially internalizing the sales stage can be relatively passive, Chinese firms rapidly organized sales networks and successfully increased the market share especially in small and medium-sized local cities and in rural areas. Handsets sold in China had been expensive and mainly for business use during the 1990s; however, indigenous firms sold entry models through their own distribution channels and expanded sales all over the country. As a result, whereas the handset markets of major cities were still largely dominated by foreign firms at that time, new subscribers welcomed the simple and low-cost handsets provided by Chinese firms. In other words, indigenous firms established their own presence in the Chinese market by creating local and rural low- and middle-end markets. The market creation was the biggest result of the home advantage at the second phase.

4.4 Analysis of the Third Phase

4.4.1 Business Environment Changes and Technology Gap

In the third phase, the business environment's changes increasingly required indigenous firms to differentiate their products. This requirement confronted them with the problem of the lack of technology again.

The changes were caused by the following four trends. The first two changes increased competitive pressures to indigenous firms. First, foreign firms changed their strategies by developing new product lines that included low-end handsets. Since major foreign firms were selling numerous handsets worldwide, therefore they could enjoy scale economies and decrease the average cost of handset significantly. In addition, they also organized part of their distribution channels, thus gaining access to the fast-growing local and rural markets. Since foreign firms possessed higher reliability of earning growth potential, it helped foreign firms build relationship with distributors and gain support. Second, numerous new indigenous firms entered the handset market following the easing and eventual cancellation of the industrial policy. In addition, a large number of illegal handsets also began to be sold in response to the rapid market growth. In China, the problem became known as "*hei shouji*" ("black handsets") at first, and then, "*shanzhai ji*" ("bandit handsets"). These handsets began to be easily made and sold under the changes of business environment, such as the vertical specialization and modularization of products. Consequently, the favorable environment for indigenous "legal" firms also became the nest of indigenous illegal firms.

Next, the latter two changes weakened the advantage in sales channels. First, as replacement demand gradually increased, consumers' needs also began to evolve. After using simple entry models, users began to demand multifunctional models. Second, chain retail stores (Gome, Suning, and so on) and telecommunication carriers also entered the distribution business, especially in the urban market, resulting in diversified distribution channels. Consequently, indigenous firms lost their advantages in the sales stage and were finally saddled with an excess of the nationwide sales networks.

These changes in the business environment implied that the focus of competition among the all players shifted to product differentiation. Because each firm extended its product range in an attempt to improve its brand recognition, almost 600 models were launched annually. Consequently, the product lifecycle sharply became shorter, and the average volume of shipments also dramatically decreased. Then, the price range had expanded from low-end products retailing at 1,500 RMB and below; to middle-end products at 1,500–2,500 RMB; to high-end handsets at 2,500 RMB and above. Models produced by indigenous firms were concentrated in the 1,000–2,000 RMB price range between the low- and middle-ranges. In this category, fierce competition pressured firms to develop differentiated models and to survive. As new handset concepts were developed and launched by foreign firms; indigenous firms competed to absorb the new trends of products and to produce low-price local versions of the foreign cutting-edge models.

The differentiation requirement, however, brought indigenous firms face-to-face with the technology gap again. Most indigenous firms had concentrated their energy on the marketing-oriented strategy since they entered; therefore they accumulated little experience in technology acquisition. Consequently, post-2003, the significant gap in experience between indigenous and foreign firms resulted in a stagnation of Chinese firms, with some indigenous firms leaving the industry.

4.4.2 External Technology

Many indigenous firms decided to continue relying on outside firms for product development (and in some cases manufacturing), because they did not have sufficient technological capabilities to develop competitive models. Although dependence on outside firms persisted, the partners entirely changed. The rapid expansion of the market led to the emergence of indigenous Chinese IDHs (Marukawa et al., 2006). Many indigenous manufacturers began to abandon transactions with Korean and Taiwanese OEMs/ODMs and IDHs, and instead began buying handsets from Chinese IDHs. The indigenous IDHs provided design services and offered handsets at lower prices than Korean and Taiwanese OEMs/ODMs and IDHs.

Although indigenous handset manufacturers continued to depend on external technology, there were also rationalities to outsource. First, because IDHs accepted orders from many indigenous firms at the same time, the average cost per model was generally low. If every indigenous firm developed and designed every new model independently, then the average cost could not be decreased as much. Second, because indigenous firms often depended on competitors to develop new functions and designs to make their product lineups attractive, IDHs could meet these similar needs of indigenous firms as well as save their business resources by reusing development outcomes. Thus, external technology was efficient in developing and widening product ranges for indigenous firms.

4.4.3 Internal Knowledge and Technology Acquisition

While small and medium-sized indigenous firms depended wholly on external technology, major indigenous firms began attempting to develop some new models independently. Development capabilities were also required to expand the handset businesses, though Chinese firms did not change their marketing-oriented strategy. However, some firms did not succeed in developing new models, since they were unfamiliar with the development stage. On the other hand, Lenovo expanded its market share by accumulating some of the development experiences on the basis of the marketing capabilities they developed through their successful PC business. To accumulate such experiences, Lenovo leveraged the easy-to-use platforms developed by MediaTek (MTK), a major Taiwanese chip vendor for consumer electronics.[17]

In contrast, numerous major firms suffered setbacks in their attempts to accumulate development capabilities. For example, Bird agreed on the establishment of a 50–50 joint venture (JV) with Sagem (France), and TCL merged with the handset division of Alcatel (France) to enhance its development capabilities; however, indigenous firms found it difficult to manage JVs and M&As because of their lack of experience. Amoi Electronics (Amoi) reported a significant deficit in accumulated capabilities because they could not reflect the increase of developmental cost in business performance, even though their focus was R&D activities. As a result, it was difficult even for major indigenous firms to increase their technological capabilities and improve business performance.

In response to this situation, indigenous Chinese firms and IDHs began accepting the easy-to-use MTK platforms. At first, MTK and an indigenous IDH, Longcheer, cooperated and refined the MTK platforms for the Chinese handset market (Shiu and Imai, 2010). As a result, in addition to IDHs, indigenous Chinese firms also began using the platforms for product development. Their platforms included a core hardware layer and was equipped almost all the software as enclosed with a bold line in Figure 4.7, and it dramatically eased the difficulties of product development and design. Although this simplification of product structure was achieved at the large expense of product differentiation, the acceptance rate of MTK platforms among indigenous firms surged from 13 percent in 2004 to 71 percent in 2005. In addition to the simplification, MTK provides reference designs for newcomers in the handset manufacturing industry. Reference designs provided by chip vendors are design drawings of products using relevant venders' platforms. The US' and EU's chip vendors also provide reference designs, but enough experiences and capabilities were also required

	Hardware	Software
Core layer	RF, Data processing (baseband chip, etc.)	Basic software (OS)
Middle layer	Circuit on PCB, Devices (display, camera, battery, etc.)	Communication middleware (protocol stack, etc.)
Surface layer	Outer case, Keypad	User interface (menu screen, etc.), Application software

Figure 4.7 Product structure (In the case of using the MTK platforms)
Note: As noted in Figure 4.5, the elements contained within the bold line are modularized as platforms.
Source: Author's creation based on various materials.

to develop products with their reference designs for handset manufacturers. In contrast, the MTK's reference designs showed complete component lists for product development using the MTK platforms. Consequently, many major indigenous firms, such as Bird, TCL, and Lenovo, have accepted MTK platforms to develop new models.

While other indigenous firms were stagnating, Lenovo grew by integrating a certain level of the development stage and using its own product policy. Lenovo was a pioneering user of the MTK platforms. Although the use of the MTK platforms compromised differentiation, Lenovo adopted a mix of variety of platforms to retain their competitiveness (interview at Lenovo in Xiamen, on August 27, 2007). Exploiting its development capabilities using the MTK platforms, Lenovo also seized business opportunities by launching a rapid succession of new products, in particular middle-end products, which suited the home market. For example, Lenovo changed its monochrome displays to colored ones across its entire product range in 2004, and the company launched handsets with an MPEG Audio Layer-3 (MP3) function in 2005, when the feature was being popularized among foreign firms. In addition, they had the capabilities to develop and design outer cases and circuits on PCBs, allowing them to release sophisticated design models. The business environment at the third phase required development capabilities as well as marketing capabilities.

Therefore, Chinese firms needed to exert the home advantage on the basis of some technologies. Only development capacities could realize to differentiate models, to diversify their own product portfolio, and to meet the market. However, handset firms could not depend on outside firms at this point. IDHs had no motivation of devoting their human resources entirely to the specific requirements of indigenous firms. As a result, models developed by IDHs have often been inferior in differentiation and quality. For this reason, major handset firms sought to decrease their dependency on external technology and began to develop their own models. In the case of Lenovo, 90 percent of new models have been designed by the company at this phase. The decision drew its rationality from the viewpoint of the boundaries of the firm.

4.5 Conclusion

This chapter has examined the growth of indigenous manufacturers facing a technology gap by exemplifying China's handset industry. Because of limited technological capabilities, indigenous firms depended on outside firms for product design and development, and in some cases, manufacturing. In particular, MTK platforms helped indigenous firms to grow in the market. However, at the same time, indigenous firms had focused on the sales stage, exploiting the home advantage and familiarity with the domestic market. Consequently they could expand their market share by using external technology and the home advantage in order to compensate for the technology gap.

In contrast, the unsustainability of internal knowledge also has been revealed in the third phase. Chinese handset firms had the business opportunity to expand their businesses by focusing on selling inexpensive products in the local and rural markets. However, major foreign firms learned the strengths of this strategy and recovered their market shares in the third phase. Although indigenous firms have absorbed technologies from foreign firms, foreign firms have also absorbed the advantages from indigenous firms. Moreover, since competitive products are important to maintain a strong manufacturer–distributor relationship, foreign firms were able to retain a much greater advantage than indigenous firms. Indigenous firms had to accumulate technologies while enjoying business opportunities created by using internal knowledge. Therefore, the home advantage has to be discounted to a certain extent in terms of sustainability.

Nevertheless, the fact remains that Chinese firms have attained growth by using internal knowledge. The sustainability is short, but the advantage can be a significant opportunity to enter markets. Just because it is an unsustainable strategy does not mean it has no effect. In the case of China's handset industry, indigenous firms are required to accumulate technological capabilities to develop new models with platforms provided by MTK and foreign chip vendors. If they did not accumulate technological capabilities by depending only on the past strategy, they would not be able to survive in the market. On the other hand, if they invest too many resources at the development stage, it would be a very risky choice like Amoi. Therefore, we can find that it is significantly difficult for firms to find optimal balances among the three factors at every stage in growth processes. The most important fact is that indigenous firms need to accumulate some acquired advantages through competition in domestic markets after entering with their inherent advantage.

CHAPTER 5

MODEL

5.1 INTRODUCTION

In this chapter, we develop a model of the diversification mechanism presented in the previous chapter to generalize the experience in China's mobile phone handset industry.[1] We have in detail analyzed the make-or-buy decisions of Chinese handset manufacturers through a decade. The case study can help us understand the complicated relationship among the three factors: the technology gap, external technology, and internal knowledge. But it is still difficult to expect that an exact optimal balance among the three factors would be established. Therefore, we develop the diversification mechanism model and identify entry conditions of indigenous firms facing a technology gap.

To do it, we employ a model of the boundaries of the firm developed by Antràs and Helpman (2004) and Antràs (2005) (the AH model). They applied the Grossman–Hart–Moore (GHM) model to analyze multinationalization of firms through international trade and outward foreign direct investment (FDI). We use the AH model and analyze the behavior of indigenous firms in developing countries in globalization. Antràs and Helpman (2004) showed that firms in developed countries can have four types of relationships with manufacturers in developing or developed countries, which as listed in descending order per the productivity of firms in developed countries: integrating manufacturers in developing countries, outsourcing to manufacturers in developing countries, integrating manufacturers in developed countries, and outsourcing to manufacturers in developed countries. Alternatively, they can be called FDI, trade, domestic investment, and domestic procurement, respectively. Antràs (2005) showed that firms in developed countries choose to manufacture in developing countries when technologies become obsolete and/or the wage in developing countries is low, even if firms in developed courtiers face the problem of incomplete contracts in developing countries. In this way, they modeled the boundary selection; however, in the AH model, they focused on the behavior of firms in developed countries and the relationship of division of labor between firms in developed and developing countries. Consequently, the AH model based on the viewpoint of firms in developed countries, not on that of firms in developing countries, which compete with foreign firms in developed countries in the same market.

Therefore, we apply the AH model to analyze firms in developing countries in globalization. Although we directly use the AH model's basic settings for the demand function and the revenue and profit functions of firms as shown in the next section, the relationship between firms in developed and developing countries are entirely different from the AH model. Unlike the AH model, indigenous firms in our model are competitors of foreign firms in markets in developing countries. Therefore, they are not mere factories to be integrated or outsourced by foreign firms in developed countries. Moreover, we are focusing on the boundary selection of indigenous firms in developing countries, and not that of firms in developed ones. In addition, to elucidate the difference between firms in developed and developing countries, the existence of the technology gap is incorporated into the model adopted by the present study.

The technology gap is considered a result of insufficient experience. When a product structure is complicated, and development skills cannot be manualized, it becomes difficult to diffuse technologies. Consequently, learning-by-doing (LBD) is required to successfully develop and design new models. In fact, in the 1990s, the nationalization project for the handset industry faced the problem of insufficient experience (Chapter 4). Since indigenous firms lacked technological capabilities, the cost to make a model increased significantly (Economic Structure Reform and Economic Operation Office of Ministry of Information Industry of China, 2003).

The following can be expected from this analysis. Even if investment is required in specific human capital to increase product value, it is rational for firms to refrain from integrating such intermediate input when the investment in human capital does not produce immediate results. This is because firms without enough experiences face relatively high average costs of products. The boundary selection of indigenous firms in developing countries facing a technology gap is thus investigated and shows the entry condition of indigenous firms.

The remainder of the chapter is organized as follows. In the next section, an analytical framework is shown. In Section 5.3, a model is developed and the entry condition is analyzed. Findings are presented in the concluding section.

5.2 MODEL

An economy in which there exists a developed country (North) and a developing one (South) is considered as the AH model did. Suppose that there is a North firm (N) and a South firm (S) in each country and that both firms are competing in the South market. The North firm enters the South market, by exporting to or investing in the South.[2] In addition, suppose that both firms input only labor and produce goods for final consumption (y).

Consumers have a simple demand function for the final goods as follows:

$$y = \lambda p^{-1/(1-\alpha)}, \ 0 < \alpha < 1, \tag{5.1}$$

where $\lambda > 0$ is a coefficient given exogenously, p is the price of the final goods and $1/(1-\alpha)$ is the price elasticity of demand.

Behaviors of firms are set here. Suppose that both firms run businesses by combining a headquarters service (x_h) and a technology service (x_t). This study defines the headquarters service as various activities conducted to explore product concepts. If product concepts are not yet fixed, because the market is still in its infancy, then the importance of the headquarters service will increase to plan new models. The technology service is defined as activities to develop and manufacture key components that make up the final product.[3] When product concepts are fixed, the importance of the technology service will increase thereby enhancing the value of the final goods by increasing the function of the key components. The headquarters service is only provided by assemblers of the final goods both in the North and South. The technology service is provided by the North firm that produces the final goods in the North and by the South firm or a component supplier in the South. The North firm provides the technology service, and the South firm may also provide the technology service by integrating the supplier, or buy it from the outside independent supplier rather than integrating the stage. This next sections examine the optimal selections of the make-or-buy decision made the South firm regarding the technology service.

5.2.1 Behavior of the North Firm

The North firm independently provides the technology service by itself. Therefore, two inputs, x_h and x_t, are combined on the basis of the Cobb–Douglas production function to produce the final goods:

$$y = \sigma_z x_h^{1-z} x_t^z, \ 0 < z < 1, \tag{5.2}$$

where z is the elasticity of production of the technology service; $\sigma_z = z^{-z}(1-z)^{-(1-z)}$. The final goods industry becomes a technology service-intensive industry when $z > 1/2$. It becomes a headquarters service-intensive one when $z < 1/2$.[4] The following revenue function of the North firm (R^N) can be derived from Equations (5.1) and (5.2):

$$R^N = \lambda^{1-\alpha} \sigma_z^\alpha x_h^{\alpha(1-z)} x_t^{\alpha z}.$$

When it bears a wage rate (w^N) in the North to produce every unit of production, the North firm chooses x_h and x_t to maximize the following profit function:

$$\pi^N = \lambda^{1-\alpha} \sigma_z^\alpha x_h^{\alpha(1-z)} x_t^{\alpha z} - w^N x_h - w^N x_t. \qquad (5.3)$$

5.2.2 Behavior of the South Firm

Unlike the North firm, the South firm does not independently provide the technology service from the very beginning first. Therefore, the South firm makes a decision of boundary selection decision, $k \in (M, B)$, that is, whether the firm makes (M) or buys (B). If the GHM model is follows, the South firm will integrate the supplier and independently make x_t in order to avoid facing the hold-up problem, particularly when the firm needs to invest in human capital so that x_t increases the value of the final goods.

Next, the influence, $\beta_k \in (0, 1)$, on gains between the South firm and the supplier is explained. Regarding shares between the two firms, suppose that both sides can receive the gains of each outside option and half of the rest, based on the Nash bargaining solution. The outside option is a gain that each side can receive when negotiations fail. If bargaining fails when the South firm buys x_t (i.e., the South firm does not integrate the supplier), then the outside option of the South firm for the technology service is zero. In contrast, even if bargaining fails when the South firm provides the technology service, the firm can receive δ as an outside option. Suppose $0 < \delta < 1$, then the South firm that integrates the technology service can keep δ^α against sales. In the case that the South firm makes x_t, the share is R^S and it is the total amount of sales in the South. This shows that the supplier has become a part of the South firm as a department of production of the technology service. In other words, the more an independent firm makes, the more their shares decrease. When the South firm buys x_t, the supplier can maintain the outside opportunity of the technology service as an independent supplier. In summary, the relationship between the two sides is as follows:

$$\beta_M = \delta^\alpha + \frac{1}{2}(1-\delta^\alpha) > \frac{1}{2}(1-\delta^\alpha) = \beta_B.$$

While the above is based on a general mechanism of the boundary selection, the influence of the technology gap on the mechanism can be connected. It is possible that a high technological level is required to produce the technology service because of the technological difficulties faced in the production of key components for the final goods. Therefore, the South firm must accumulate experiences to master technologies by itself. However, the South firm's productivity was initially lower than that of the North firm due to the lack of experience. Hence, even if the South firm integrates the supplier to improve the quality of the key components, it will not receive the same effects of human capital investment as the North firm. Consequently, it can be assumed that the average cost in the South (a_k) increases by $a_M > 1$ when the South firm chooses to produce the technology service in-house, even though the wage rate in the North is higher than a wage rate (w^S) in the South. We define w^N as more than twice the value of w^S, $w^N/2 > w^S$. Therefore, while the wage rate in the South is lower than that in the North, depending on the size of the technology gap, the labor cost in the South possibly exceeds that in the North.

Hence, the South firm's and the supplier's revenues are, respectively, as follows:

$$\begin{aligned} R_f^S &= \beta_k \lambda^{1-\alpha} \sigma_z^\alpha x_h^{\alpha(1-z)} x_t^{\alpha z}, \\ R_s^S &= (1-\beta_k) \lambda^{1-\alpha} \sigma_z^\alpha x_h^{\alpha(1-z)} x_t^{\alpha z}. \end{aligned} \quad (5.4)$$

When the South firm provides the technology service, the supplier receives the above revenue as part of the South firm. Next, the South firm's and the supplier's profits are, respectively, as follows:

$$\pi_f^S = \beta_k \lambda^{1-\alpha} \sigma_z^\alpha x_h^{\alpha(1-z)} x_t^{\alpha z} - \alpha_k w^S x_h,$$
$$\pi_s^S = (1 - \beta_k) \lambda^{1-\alpha} \sigma_z^\alpha x_h^{\alpha(1-z)} x_t^{\alpha z} - w^S x_t. \tag{5.5}$$

The equation is similar to Equation (5.4). When the South firm provides the technology service, the supplier receives the above profit as part of the South firm. In addition, it can be assumed that the supplier also employs workers at lower wage rates than those offered by the North firm. This is because the South firm and the supplier are located in the South irrespective of integration or disintegration by the South firm.

5.3 Equilibrium

In this section, optimal prices and marginal costs for the North and South firms, derived from the profit functions in the previous section, are developed. In addition, the boundary selection of the South firm is analyzed. Certain possibilities to mitigate the entry condition are also considered.

From Equation (5.3), the optimal price for the North firm is as follows:

$$p^N = \frac{w^N}{\alpha}. \tag{5.6}$$

The price depends on the wage rate and price elasticity of demand in the North.

Next, from Equation (5.5), the optimal price for the South firm is as follows:[5]

$$p^S(\beta_k) = \frac{\alpha_k^{1-z} w^S}{\alpha \beta_k^{1-z}(1-\beta_k)^z}. \tag{5.7}$$

The North firm does not need to make a decision regarding the boundary selection. Hence the optimal price depends on the wage rate in the North. In contrast, the South firm is required to choose boundaries that minimize the optimal price. Then, the South firm must choose β_k depending on z and on the influence of $a_M > 1$ for the optimal price when option to integrate.

We first consider the influence of z on the boundary selection, $k \in (M, B)$. The South firm chooses β_k to minimize the optimal price under a certain z. Examining β_k in Equation (5.7) reveals that the South firm can minimize the optimal price when it chooses a smaller β_k if z is bigger, and conversely, the larger one if z is smaller. In cases where the industry is the technology service-intensive one, the supplier's investment in human capital increases more than that of the South firm. This is because the key component is of higher significance to the product value. In contrast, if the industry is the headquarters service-intensive one, the South firm's investment increases more than of the supplier. This is because the headquarters service is of higher significance to the product value. Therefore, buying is the optimal boundary when it is a technology service-intensive industry; in contrast, making is optimal when it is a headquarters service-intensive industry.

Next, the entry condition for the South firm can be set based on optimal prices. Here we assume that the North and South firms are competing in the South market under the Bertrand competition model.[6] Consequently, the South firm's optimal price should be lower than that of the North because rational consumers will not buy homogeneous goods at higher prices. It is unrealistic to expect that both the North and South firms, that is, foreign and indigenous firms, make homogeneous goods and compete in the same market. This is because the South firm tends to avoid direct competition with a North firm in the same market. However, to explicitly consider the influence of a North firm on a South firm's boundary selection, direct competition between both the North and

South firms is assumed here. In this case, under the Bertrand competition, the entry condition becomes $p^N/p^S > 1$.

However, it is not adequate to make a simple comparison between the optimal prices of North and South firms to assess the entry condition. Since the optimal prices for both the firms are monopolistic, there is room for price reduction in response to another competitor for both parties. This can result in our model having no equilibrium. To avoid the influence of monopoly prices and compare true competitiveness of North and South firms, we should compare the lowest prices, that is, marginal costs (MC), of both as the actual entry condition. In our model, MC can be obtained by multiplying the monopoly prices by α.[7] Since the monopoly prices of North and South firms are Equations (5.6) and (5.7), respectively, therefore the marginal costs of them are as follows, respectively:

$$MC^N = w^N, \qquad (5.8)$$

$$MC^S(\beta_k) = \frac{\alpha_k^{1-z} w^S}{\beta_k^{1-z}(1-\beta_k)^z}. \qquad (5.9)$$

To ensure to compare the marginal costs as the entry condition, we suppose that there are two homogeneous North firms in the South market and one South firm that is seeking to enter the market. This is because we suppose the two North firms is to equalize their prices at the marginal cost of the North firms under the Bertrand competition. Therefore, if the marginal costs of the South firm are higher than those of the North firms, then the South firm cannot enter and the marginal costs of the North firms become the equilibrium price in the market. In contrast, if the marginal costs of the South firms are lower than those of the North firms, then the South firm, in turn, enters and a price that is slightly lower than the marginal costs of the North firms becomes the equilibrium one in the market.

Under the model setting and the assumptions, the following condition from Equations (5.8) and (5.9) arises:

$$\omega \geq A(\beta_k), \tag{5.10}$$

where ω and $A(\beta_k)$ are defined as follows:

$$\omega = \frac{w^N}{w^S},$$

$$A(\beta_k) \equiv \frac{\alpha_k^{1-z}}{\beta_k^{1-z}(1-\beta_k)^z}.$$

Thus, the South firm must choose boundaries in which $A(\beta_k)$ is equal to or smaller than ω. Depending on circumstances, the South firm may not be able to enter the market.

5.3.1 Technology Service-Intensive Industry

Based on the entry condition, the South firm's boundary selection in the technology service-intensive industry can be considered. In the case of the technology service-intensive industry, the South firm can run its business by purchasing key components. The condition is as follows:

$$\omega = \frac{1}{\beta_B^{1-z}(1-\beta_B)^z}. \tag{5.11}$$

As shown in Equation (5.11), the South firm chooses to buy, $a_B = 1$, in the technology service-intensive industry. The South firm does not need to incur the increased production cost caused by making the technology service. Although the right-hand side of Equation (5.11) is greater than 1 because $\beta_B < 1/2$ and $1/2 < z < 1$, the South firm can enter because $w^N/2 > w^S$.

5.3.2 Headquarters Service-Intensive Industry

Next, the South firm's boundary selection in the headquarters service-intensive industry is considered. The firm chooses among making, buying, and nonentry depending on the wage ratio ω and a_k. This condition is observed in Equation (5.10). Even if the industry is headquarters service-intensive, the firm is required to choose boundaries in consideration of the burden of $a_M > 1$ when the South firm independently makes. Therefore, the South firm can fulfill the entry condition if the wage ratio ω is large enough to cancel out the burden. Under this condition, the South firm integrates the supplier.

When the South firm has to bear excessive costs, $a_M > 1$, even if integration is optimal, then there is a possibility that the South firm can clear the entry condition by choosing to buy. Because the South firm has to bear $a_M > 1$, even if integration ($\beta_M > 1/2$) is optimal in the headquarters service-intensive industry ($z < 1/2$), the South firm can decrease a_M to $a_B = 1$ by choosing to buy ($\beta_B < 1/2$). Consequently, if the South firms maintains $A(\beta_M) < \omega$, it can satisfy the entry condition. The new entry condition is similar to that in Equation (5.11). However, $A(\beta_M)$ increases in comparison to the boundary selection, which is optimal when choosing to buy because the South firm chooses $\beta_B < 1/2$ despite the fact that $z < 1/2$. When the South firm cannot fulfill the new "compromising" condition, the firm chooses nonentry, in other words, zero boundaries. Therefore, in the headquarters service-intensive industry, the boundary selection can be more complicated than the selection in the technology service intensive industry. There are three patterns: making, buying despite the fact that making is an optimal selection, and nonentry.

5.3.3 Mitigation of Entry Conditions

Finally, this evaluation considers two cases of easing the entry condition. In the first case, when the final goods become products of the technology service-intensive industry as shown in Equation (5.11), the South firms do not need to incur the burden of $a_M > 1$. Therefore, if technologies change and z in Equation (5.2) increases, then the final goods are changed from products of the headquarters service-intensive industry to those of the technology service-intensive industry. Even if South firms cannot enter when the industry is the headquarters service-intensive insudstry, the possibility for them to enter will increase.

This case can be found in reality. As shown in the previous chapters, Chinese electronics firms have gained business opportunities by depending on external technology. Owing to the vertical specialization and the modularization of product structures, they have not been required to independently develop and manufacture key components. If indigenous firms needed to make them, it would be far more difficult for indigenous firms to enter the markets.

The second case involves the home advantage. Foreign firms have a technological advantage in the market; however, it is possible that they cannot exert the effects of the headquarters service in a given market without understanding consumer preference and adapting to its business practices. This is the away disadvantage. Therefore, as the North firm has the advantage in production costs for the technology service, it is possibility that the South firm has the home advantage in production costs for the headquarters service used in the South market. The South firm can then increase revenue R_f^S according to the degree of the advantage. Thus, the South firm can choose boundaries to offset the technology disadvantage using the home advantage. Suppose that the increment of the revenue is $\varphi > 1$, then Equation (5.4) will be as follows:

$$R_f^S = \beta_k \varphi \lambda^{1-\alpha} \sigma_z^\alpha x_h^{\alpha(1-z)} x_t^{\alpha z},$$
$$R_s^S = (1 - \beta_k) \lambda^{1-\alpha} \sigma_z^\alpha x_h^{\alpha(1-z)} x_t^{\alpha z}.$$

The optimal price of the South firm decreases by φ^{1-z}, which is less than that in Equation (5.7), and the entry condition is mitigated follows:

$$p^S(\beta_k) = \frac{\alpha_k^{1-z} w^S}{\varphi^{1-z} \beta_k^{1-z}(1-\beta_k)^z}.$$

This case can also be found in reality. Chinese electronics firms have been achieving growth by using internal knowledge. In fact, they have identified business opportunities much earlier than foreign firms and increased their presence in the Chinese market. Therefore, if indigenous firms did not have the home advantage, foreign firms would have dominated the fast-growing Chinese market.

5.4 CONCLUSION

Based on the AH model, this chapter has shown that the boundaries of a South firm, that is, indigenous firms, in globalization can be diversified in comparison with those of a North firm, that is, foreign firms. Specifically, we have analyzed whether or not indigenous firms make technologically advanced goods in-house when faced with a technology gap. It is possible to generalize our major conclusions by extracting the entry conditions. At first, we have investigated the cases of the technology service- and headquarters service-intensive industries. Incorporating the technology gap revealed that the indigenous firms chose to buy if the technology gap was significantly large and if they depended on the supplier; the latter gave different results for the firm's boundary selection theory. We have explicitly analyzed the influence of the existence of productive foreign firms on the boundary selection of indigenous firms.

On that basis, we have replicated the experiences of Chinese electronics firms, which have been discussed in the previous chapter. The entry condition of indigenous firms can be eased in the technology service-intensive industry and when indigenous firms exert the home advantage. Consequently, indigenous firms can enter and grow using external technology and/or internal knowledge, even when faicng the technology gap.

In addition, the cases of homogenization of the boundaries of indigenous firms and of zero boundaries have also been shown, although the case of diversification has been mainly focused in this chapter. As a result, we have got to be able to relativize and predict the three cases by considering the three factors, given that the wage ratio is sufficiently large. From a practical perspective, the boundaries of indigenous firms can occur as diversified ones between both ends of homogenization and zero boundaries.

CHAPTER 6

CHALLENGE FOR OVERSEAS EXPANSION

6.1 INTRODUCTION

Outward foreign direct investment (FDI) from developing countries is increasing.[1] According to UNCTAD (2001, 2012), the ratio of outward FDI from developing countries to total outward FDI in the world was just 8.7 percent in 2000, but rapidly rose to 26.9 percent in 2011. Traditionally, FDI investors came exclusively from developed countries, but this scenario changed radically in just a decade—at present a sizable number of FDI investors come from developing countries. This has made it even more important for us to understand the characteristics of outward FDI from developing countries at the moment.

Among the developing countries, China has risen as a major investor. China's manufacturing sector has been developing rapidly since the 1980s. Thus, since the mid-1990s, Chinese firms have faced the need to identify various factors to further their growth, such as expanding sales in overseas emerging markets, developing natural resources, and enhancing their research and development (R&D) capabilities. Consequently they began establishing sales offices and factories in foreign countries to establish their presence in overseas markets, forming joint ventures (JVs) with indigenous mining firms in host countries in order to acquire natural resources, and setting up laboratories in developed countries to increase their R&D capabilities.[2] To encourage the outward FDI of Chinese firms, the Chinese government has been implemented the "Going Global" (*Zouchuqu*) policy since the late 1990s. This study focuses on China's outward FDI for expanding sales in order to investigate how Chinese firms seek to realize further growth amidst fierce competition in the global market.

Previous theoretical studies on outward FDI have primarily investigated the determinants of outward FDI. One of the main propositions in this research field is that only competitive and productive firms can invest in foreign countries and become multinational enterprises (MNEs) (Antràs and Helpman, 2004; Dunning and Lundan, 2008; Helpman et al., 2004). They identified that competitiveness such as advanced technologies, management know-how, strong brand recognition, high productivities, and so on, are determinants for outward FDI.[3] This is because outward FDI has both advantages and disadvantages for investing firms. Although investors can increase their sales in host countries by investing, they must incur additional costs to sell their products in unfamiliar markets. More precisely, they must be able to tackle the cost and time needed to learn consumers' preferences, information and business practices in host countries, and so on. This book has referred to this phenomenon as the away disadvantage. Therefore, MNEs must be competitive and productive to bear the additional costs.[4]

To this effect, MNEs can be considered rather successful in their home countries in comparison to noninvestors in the same countries. However, this does not hold true for all MNEs, that is, not every MNE is equally competitive and productive in the host country. In particular, successful firms from developing countries are not as competitive and productive as those from developed countries and, in some cases, even those in host countries. However, previous studies on outward FDI only elucidate the determinants of outward FDI from home countries, and not the feasibility or potential of outward FDI from developing countries to foreign ones. Therefore, we aim to find a condition that MNEs from developing countries can survive when competing with more aggressive and competitive MNEs from developed countries as well as against indigenous firms in host countries.

Since we have shown that Chinese firms have realized growth with the home advantage as local firms, there is a possibility for firms to be unable to exert the advantage when they are separated from domestic markets. In other words, Chinese investors rather turn to face the away disadvantage. In addition, indigenous firms in other developing countries may also have access to external technology due to the development of the vertical specialization, although well-developed industrial agglomerations may be required to purchase various components efficiently. In fact, even major Chinese electronics firms often have small shares in overseas markets. Neverthelss, Chinese firms can also exert their advantage in overseas markets and develop competitiveness, although they have not yet dominated all over the global market. To this end, certain previous studies on outward FDI from China have focused on the aspect of weakness as firms from developing countries.[5] Williamson and Zeng (2009) showed that Chinese MNEs face some weaknesses, such as brand recognition, the lack of technology, and so on. Specifically, Yuan (2013) showed that Chinese automotive and electronics MNEs in ASEAN lack competitiveness in comparison to MNEs from Japan and South Korea (Korea). Naturally, MNEs from developing countries do face weaknesses when competing in the global market; thus, measures must be identified to alleviate these weaknesses. Therefore, we will focus on relationships with firms in host countries, which, to the best of the author's knowledge, have not been considered in studies.

To identify the condition influencing transactions with indigenous firms in host countries, this study examines Chinese MNEs that manufacturer television (TV) sets in South Africa (SA) and conducted an interview-based case study. This is because the sector can be a typical example of MNEs from developing countries.[6, 7] Chinese TV set manufacturers have used low-wage workers and produced price-competitive products in China. While they have developed a high degree of assembling capabilities, they have not been able to accumulate a similar level of capabilities in R&D activities. In the 1990s as competition tightened in the Chinese market, some Chinese manufacturers began exporting and investing in overseas markets. Meanwhile, Chinese manufacturers began targeting the SA market as an investment location for the following reasons. First, the SA market had begun to grow rapidly after the end of apartheid in the early 1990s (Chen, 1994; Wu, 2005).[8] The SA government was democratized, eliminating racial discrimination. As a result, the income of black people rose, and many foreign firms such as Chinese firms decided to enter the market and capture some of the growth. The second reason was that SA and China began having economic exchanges in the early 1990s. The two countries did not have diplomatic relationships because the Communist Party of China had built ties with the African National Congress, led by the late and former president Nelson Mandela, which opposed the white government in SA. It was only following the transition to the black leadership and democratization that the two countries started to have economic transactions, and finally in 1998, established diplomatic ties (Qian, 2003; Yang, 2010). The change paved the way for Chinese firms to export to and invest in SA.

At present, Chinese manufacturers in SA are still not as competitive and productive as MNEs from developed countries and indigenous SA firms. To compensate for their weakness, they have been partially producing and selling products under the brand names of indigenous SA firms. They do not have the power of brand recognition or the R&D capabilities to differentiate their products enough; therefore they have depended on the brand recognition and sales channels of indigenous SA firms for sales expansion. We will investigate the strategies of Chinese MNEs to exemplify firms from developing countries and will seek the condition required for their survival amidst competition in foreign markets. Of course, although there are numerous studies on the outward FDI from China, this study will focus on the aspect of firms in developing countries, especially on the transactions with indigenous firms in the host country.[9]

This chapter is organized as follows. Section 6.2 provides an overview of Chinese firms in SA and their tariff-jumping investments. Section 6.3 examines the business methods of Chinese manufacturers in SA. The final section sets out this study's conclusions.

6.2 CHINESE FIRMS IN SOUTH AFRICA AND TARIFF-JUMPING INVESTMENTS

6.2.1 OVERVIEW OF CHINESE FIRMS IN SOUTH AFRICA

Immediately after economic exchanges commenced between China and SA in the early 1990s, some Chinese firms took the opportunity to invest in SA and built factories in order to expand sales. Table 6.1 lists Chinese companies that have invested in SA's home appliances manufacturing industry. The first investor was SVA Group (SVA). SVA is one of Shanghai's state-owned enterprises (SOEs). It set up operations in SA in 1993 and has produced TV sets, flat panel displays (FPDs) for TV sets, and various home appliance and consumer electronics products. However, in 2009, SVA suffered financial problems and was taken over by Shanghai Yidian Holding, which is also an SOE of Shanghai, to rebuild its business. When SVA entered the SA market, they built a factory near Johannesburg in 1993 to produce TV sets with black-and-white cathode-ray tubes (CRTs) (*Lingdao Xinxi Juece*, No. 26, 2008). The firm produces various white goods of home appliance products at present. They have 500 workers employed at the factory as of 2010 (interview at the Ministry of Commerce of the People's Republic of China in Beijing, on June 25, 2010).

Table 6.1 Chinese home appliance manufacturers in SA

Name	Headquarters	Entry year	Mode	Products	Remarks
SVA	SOE of Shanghai	1993	Local production (1993–present)	White goods (e.g., refrigerators, washing machines), TV sets and other electronic products	500 workers
Hisense	SOE of Qingdao, Shandong	1993	Export → Local production (1997–present)	TV sets and other electronic products	230 workers. Purchased the SA factory of Daewoo (Korea) in 2000
XOCECO	SOE of Xiamen, Fujian	1998	Local production (?–present)	TV sets, DVD players and other electronic products	130 workers

Sources: Interviews at the Ministry of Commerce of the People's Republic of China in Beijing (June 25, 2010); Konka in Shenzhen (December 2, 2010); Professor Stephen Gelb (University of Johannesburg) in Johannesburg (September 5, 2011); HiFi Corp in Johannesburg (September 10, 2010; September 9, 2011); Hisense in Qingdao (October 28, 2011). Also Gelb (2010), Wang and Ding (2012) and other literature.

The second Chinese investor was Hisense Group (Hisense). Hisense is an SOE from Qingdao Shangdong. Its predecessor was established in 1969 as a firm that produced radios. They expanded their product lineup into home appliances and consumer electronics, especially in the 1980s, and have become a major Chinese manufacturer. In the course of its rapid development, the firm entered SA in 1993. They first chose to export their home appliances from China to SA to evaluate the potential of the SA market. Once that were confident of the potential for growth, the firm set up Hisense SA Development Enterprise along with a factory near Johannesburg in 1997. It mainly produces TV sets.

The third investor was Xiamen Overseas Chinese Electronic (XOCECO). Although they were an SOE of Xiamen,

Fujian, a Taiwanese firm is presently the biggest shareholder. They were established in 1985 as a major manufacturer and exporter of TV sets. They established Sinoprima Investment & Manufacturing SA near Johannesburg in 1998 to expand sales in SA.[10] They also have a factory at the site. They have produced TV sets and Digital Versatile Disc (DVD) players, and so on. They employed 130 workers as of 2010 (interview at the Ministry of Commerce of People's Republic of China in Beijing, on June 25, 2010). Although Skyworth-RGB acquired the subsidiary of XOCECO in June 2014 to expand their SA business, we limit the discussion before 2011.

As mentioned above, these three investors are major Chinese manufacturers or champions in China. Table 6.2 is an abbreviated list of the top 100 firms in China's home appliance and information and communication technology (ICT) industries as of 2011. The 100 firms were ranked by the Ministry of Industry and Information Technology (MIIT) on the basis of a comprehensive evaluation that included sales, profits, R&D expenditures, and so on. The three investors, SVA, Hisense, and XOCECO, are ranked Nos. 32, 6, and 84, respectively.

In addition to the three investors, other firms have also been operating businesses in SA to expand sales. For example, Huawei Technologies (Huawei), a global telecommunication equipment provider, sells telecommunication equipment mainly to telecommunication carriers in SA. It has sales offices and technical centers in SA (interview at Huawei in Johannesburg, on September 10, 2010). Lenovo, a global personal computer (PC) maker, sells PCs in the SA market. Konka Group (Konka) exports TV set components to Tedelex, an indigenous SA manufacturer. TCL exported TV set components to an indigenous SA manufacturer in the 2000s, although the current status on the exports is unknown.[11] Konka and TCL are also major home appliance and consumer electronics manufacturers. From this discussion it can be seen that many Chinese firms have entered SA to expand their sales in its growing market.

Table 6.2 The top 100 firms in China's home appliance and ICT industry, 2011 (Abbreviated)

Rank	Name
1	Huawei
2	Lenovo
3	Haier
4	Great Wall Technology
5	ZTE
6	**Hisense**
7	Changhong
8	TCL
9	Founder
10	BYD
11	Panda
12	Jinglong
13	Inspur
14	Skyworth-RGB
15	Tongfang
16	Alcatel-Lucent
17	Konka
⋮	
32	**SVA**
⋮	
84	**XOCECO**
⋮	

Source: Author's creation based on an announcement from Ministry of Industry and Information Technology (MIIT).

6.2.2 Tariff-Jumping Investments

The reason underpinning SVA, XOCECO, and Hisense's establishment of factories in SA is the country's high import tariffs on TV sets. Table 6.3 depicts the tariff rates of TV sets (finished products). When firms in countries other than the European Union (EU), the European Free Trade Area (EFTA), and the Southern African Development Community (SADC) export TV sets to SA, an import tariff of 25 percent

Table 6.3 SA's import tariff rates, 2011 (%)

Classification Code	Item	Tarriff Rate			
		General	EU	EFTA	SADC
8528.7	TV				
8528.72	Color				
8528.72.20	CRT	25	3.25	13	Free
8528.72.40	Other, with a screen with no side exceeding 45 cm	Free	Free	Free	Free
8528.72.50	Other, with a screen size exceeding 3 m × 4 m	Free	Free	Free	Free
8528.72.90	Other	25	3.25	13	Free
8528.73	Black-and-White				
8528.73.20	CRT	25	3.25	13	Free
8528.73.40	Other, with a screen with no side exceeding 45 cm	Free	Free	Free	Free
8528.73.50	Other, with a screen size exceeding 3 m × 4 m	Free	Free	Free	Free
8528.73.90	Other	25	3.25	13	Free

Source: Author's creation based on revised data from South African Revenue Service (http://www.sars.gov.za/).

is levied on finished products. TV sets are primarily classified as CRT, FPD, and black-and-white CRT types (their respective classification codes are 8528.72.20, 8528.72.90, and 8528.73.20).[12] In contrast, TV set components can be imported almost duty free. Consequently, firms have begun importing components and assembling them in SA.

By imposing high import tariffs on TV sets, the SA government has compelled firms to establish factories in the country. Although this protectionist policy has promoted local production, it has led investors to set up factories that merely assemble components, known as semi-knockdown (SKD) production.[13] SKD production is defined as the assembling of various components and printed circuit boards (PCBs), which are mounted semiconductor parts. This type of production takes place because SA's component industry has not been developed. In addition, a major reason is the imposition of high tariff rates on TV sets imported as finished goods, although parts and components can be imported almost duty free. Such production does not absorb many workers. The SVA and XOCECO factories in SA have employed only 500 and 130 workers, respectively, as shown in Table 6.1.

SKD production is not limited to Chinese firms. Dominant players in the SA TV set market such as Samsung Electronics (Samsung) and LG Electronics (LG) from Korea, and Sony from Japan, also perform assembling activities in SA. Therefore, there is a vicious circle due to SA's underdeveloped component industry, and thus far the government has not been successful in nurturing the TV set industry including the manufacturing of TV set components. The government had initiated efforts to change the scope of import tariffs so that manufacturers switch to complete-knockdown (CKD) production. However, SKD production has continued in SA because firms can import TV set components practically duty free (interviews at Konka in Shenzhen, on December 2, 2010, and at Sony in Johannesburg, on September 8, 2011). This merits further study into how the SA government will adjust its tariff and trade policies and their effects on the development of the country's home appliance industry including the component industry.

6.3 Investors' Operations in South Africa

Although the three Chinese investors in this study are champions in China, they have been confronted in SA with the problem of weak brand recognition. In general, this has been a problem for Chinese firms in overseas markets. They continue to lack sufficient R&D capabilities to differentiate their products and build brand recognition in comparison to global champions from developed countries and with indigenous champions in home markets.[14] Therefore, they buy key components from outside firms. The firms can sell their products under their own brand names in China where they have brand recognition. However, when exporting their products, more often than not, they are sold under the brand names of indigenous firms in SA.

To offset the weak brand recognition, the three investors have expanded their businesses as original equipment manufacturers (OEMs) as well as sell their products under their own brand names. OEMs are firms that produce products for buyers, which are sold under the buyers' brand names. For the Chinese firms operating in SA, buyers can be divided into two groups: indigenous SA manufacturers with their own brand names and large indigenous SA retail chains that have their own store brands (SBs).[15] The former includes firms such as Defy, Tedelex, and AMAP, and so on. Although they are not well known in the global market, they are major indigenous SA firms that have SA consumers' brand loyalty. The latter group includes Pick n Pay Stores (Pick n Pay), Game Stores (Game), and so on. For example, Pick n Pay and Game have their own SBs, AIM, and LOGIC, respectively. Both have their own large sales channels across SA; therefore the Chinese investors can sell numerous products through them. The buyers also have the advantage of buying low- and middle-end products, allowing them to expand their product lineup. This is very important when targeting the growing market of black people in SA. Through the OEM businesses, the three Chinese investors have been able to expand their sales in SA despite the weak brand recognition.

Table 6.4 lists the OEM businesses in SA of the three Chinese investors, which the author was able to verify. SVA makes products for Defy and Pick n Pay. Defy is a well-known white goods manufacturer in SA, and the products that SVA makes for the firm are sold under the brand name, Defy. Pick n Pay is one of SA's large retail chains, and SVA's products are sold under its SB, AIM. Hisense made products

Table 6.4 OEM businesses of the Chinese investors

Name	Buyer	Remarks
SVA	Defy (manufacturer), Pick n Pay (retailer)	
Hisense	A manufacturer using the brand name of Sansui	No OEM business in 2011
XOCECO	Game (retailer)	

Sources: Interviews at The Ministry of Commerce of the People's Republic of China in Beijing (June 25, 2010); Konka in Shenzhen (December 2, 2010); Professor Stephen Gelb (University of Johannessburg) in Johannesburg (September 5, 2011); HiFi Corp in Johannesburg (September 10, 2010; September 9, 2011); Hisense in Qingdao (October 28, 2011). Also Gelb (2010), Wang and Ding (2012) and other literature.

for a manufacturer that sold its products under the Sansui brand; however, Hisense conducted no OEM business in 2011 (interview at Hisense in Qingdao, on October 28, 2011). XOCECO has been making products for the large retail chain, Game, which sells them under its SB, LOGIC. In this way, the three investors have been operating businesses under their own brand names and OEM production with indigenous SA firms.

Along with OEM businesses, the Chinese firms have been engaged in supplying components. This is another way that Chinese firms have been able to maintain their operations and expand sales in SA.[16] TCL and Konka (both listed in Table 6.2) export TV set components to SA manufacturers. Finished products containing these components are sold under the brand names of SA manufacturers. In the early 2000s, Konka sold CRT TV sets under its own brand name in the SA market; however, at present the firm is only a component supplier.

As indicated in the previous discussion, the share of Chinese firms operating under their own brand names is small in the SA market; however, their "production share" by firm is much larger. In 2011 the market share by brand name in SA was as follows. For CRT TV sets, 40 percent of the sets were sold by LG and Samsung. As in other overseas markets, the two Korean firms account for a large share. The remainder is shared by Telefunken (Germany), Hisense, Tedelex, XOCECO, and retailers' SBs (SA). In the LCD TV set market, 70 percent is held by Samsung, LG, and Sony. The remainder is shared by XOCECO, Hisense, Telefunken, and retailers' SBs. In the CRT and LCD TV set markets, Hisense and XOCECO have approximately 10 percent shares in the CRT and LCD TV set markets. Indigenous SA manufacturers, notably Defy and KIC, dominate the white goods market. This is because white goods are close associated with the daily lives of local consumers, and products designed by indigenous SA firms sell well; this can be the home advantage for indigenous SA firms. Two of the three Chinese investors have only a small share of the SA market. According to information on market share, apparently, SVA has none. However, through their OEM businesses, they make parts and components for white goods sold by Tedelex, Defy, and retailers' SBs.

Although Chinese firms have expanded by partially depending on the OEM businesses, they have been in direct competition with indigenous SA firms. Chinese firms have realized growth by targeting low- and middle-end markets in China. However, since indigenous SA firms are there in the host country's market, it can be said that Chinese firms have managed to maintain their businesses by operating as OEMs. In the market targeted by Chinese firms, whether Chinese investors try to sell products under their own brand names or continue operating as OEMs depends on the business strategy of each investor.

The degree of dependence on the OEM business differs by the three investors. SVA once emphasized business in its own brand names but now depends on its OEM business. XOCECO seems to be placing equal emphasis on both businesses, while Hisense's emphasis is on expanding its own brand products. The differing approaches of the three companies need further investigation to identify their determinants;

however, it appears that expanding businesses under their own brand names will be essential for further growth of the three Chinese investors. From Hisense's experience in SA, the OEM business has both advantages and disadvantages; therefore, firms have to balance the business under their own brand names with that as OEMs (interview at Hisense in Qingdao, on October 28, 2011). Although the OEM business provides the investors with opportunities to expand sales, it also has disadvantages. Their sales fluctuate with those of the buyers, and their margins are low. Since the OEM business depends on the buyer's performances, it is difficult for Chinese firms to maintain a stable expansion of their businesses. Moreover, the investors' dependence on the brand names and sales channels of the buyers weakens their bargaining power to control prices and improve profits. Thus, Hisense is attempting to promote its own brand names and not depend on the OEM business.

6.4 Conclusion

This chapter has investigated the behavior of MNEs from developing countries by examining Chinese TV set manufacturers in SA. As shown in the previous chapters, Chinese electronics firms have been growing by using external technology and internal knowledge, although they have faced the technology gap. Moreover, they cannot utilize the home advantage in foreign markets; on the other hand, they can also face the away disadvantage as foreign firms. Therefore, even champions in the domestic market can face challenges in expanding their businesses overseas if they do not possess technologies as per global knowledge.

In fact, Chinese MNEs have been competing with indigenous SA firms mainly in low- and middle-end markets. Chinese firms have attained growth by acquiring the fast-growing low- and middle-end markets in China. However, since indigenous firms in other developing countries can also employ similar growth strategies, even major Chinese firms can face difficulties in expanding their market share at first, particularly if there are indigenous firms in the host market.

Therefore, Chinese MNEs have maintained and expanded their businesses in SA by operating as OEMs to depend on brand recognition and sales channels of indigenous SA firms. As indigenous firms in developing countries can achieve growth by using the home advantage, they have possibly operate businesses by engaging in business transactions with indigenous firms in host countries in order to complement the lost home advantage.

However, it is imperative that Chinese MNEs decrease the technology gap and use an acquired home advantage, so that they can realize further growth in the global market. Chinese MNEs are required to acquire more advanced technologies to differentiate products, which can increase their brand recognition. Moreover, they must utilize their competitiveness developed under fierce competition in the Chinese market, that is, the acquired home advantage; although thus far, they have been utilizing the inherent home advantage as local firms. Indigenous firms are expected to comprehend markets in emerging countries better than MNEs from developed countries; therefore they can develop unique products for markets in the host countries. To do so, acquiring more technologies is also needed for further growth.

Conclusion

C.1 Diversification Mechanism

This study has developed the diversification mechanism for indigenous firms under competition with foreign firms. As the precondition of diversification in organization, this study has verified that technologies have not always diffused to China's electronics industry. To do it, Chapter 3 has analyzed whether or not inward foreign direct investment (FDI) has a negative impact on the growth of Chinese firms. The results have shown that inward FDI has the negative influence on sectors facing large gaps in technology and insufficient operational experience. Thus, despite rapid industrial development under globalization, the results have shown that technologies have not completely diffused to China. Therefore, indigenous firms relatively have focused more on the sales stage than the development stage to compensate for the technology gap under fierce competition with foreign firms.

On the basis of the precondition, we have explored the organizational form of indigenous firms. The make-or-buy decision of indigenous firms has been analyzed in the case study of China's mobile phone handset industry and in the model analysis in Chapters 4 and 5, respectively. The analysis has shown that indigenous firms facing a technology gap tend not to integrate stages that require advanced technologies, such as product development and design, and the manufacturing of key components. In contrast, they tend to integrate stages for which they can exert the home advantage, such as marketing in domestic markets.

Thus, the diversification mechanism of indigenous firms in developing countries has been elucidated through a case study of China's electronics industry. Drawing on the previous studies cited in Chapter 1, two natures of the diversification mechanism have been found. One is related to the literature review in Section 1.2.1. Previous studies on the influence of developed countries on developing countries have focused almost exclusively on technological homogenization and the conditions for technology diffusion. However, indigenous firms actually cannot catch up to technological levels of foreign firms in a moment completely. Consequently the organizational form of indigenous firms can be diversified by balancing the three factors of the technology gap, external technology, and internal knowledge.

The diversification of indigenous firms has been illustrated in Figure C.1. The upper thick arrows in gray, black, and white show the various types of knowledge of indigenous firms, while the lower thick arrow shows the knowledge of foreign firms. Knowledge can be divided into global knowledge and local knowledge, and they are technologies and the home advantage, respectively, as described in the Introduction. Generally, technologies can work transnationally, while the home advantage works only in domestic markets. Comparing the lengths of the thick arrows in black of indigenous and foreign firms, it shows that foreign firms have more technologies than indigenous firms. This is primarily because they started their businesses much earlier than indigenous firms did. Although indigenous firms as well are acquiring technologies by absorption from foreign firms, accumulation through production, and development through research and development (R&D) activities, they still face a technology gap between indigenous and foreign firms. However, when they can compensate for this technology gap by using external technology to add to the present technological level, and by using internal knowledge to counter the technology gap, thus they attain growth by diversifying their organizational form.

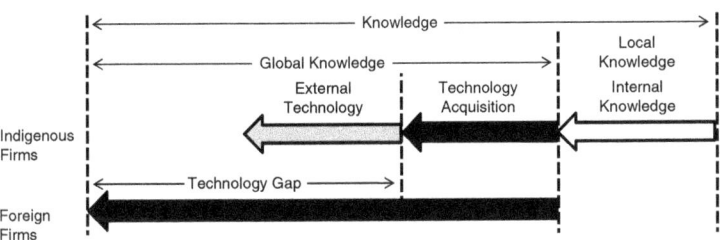

Figure C.1 Diversification mechanism
Source: Author's creation.

Another of the natures is relevant to the literature review in Section 1.2.2. Previous studies on industrial development in China have focused on the heterogenization of Chinese firms. However, since they have not explicitly investigated the relationship with foreign firms as competitors, the optimal balance of cost and benefit between heterogenization and homogenization has not been shown. Thus, drawing on the two concepts of homogenization and heterogenization, this study has shown three possibilities for the growth processes of indigenous firms in developing countries in terms of the boundaries of the firm.

The first is to select homogenization by closing the technology gap through technology acquisition. In this case, indigenous firms follow or, in a positive sense, imitate the boundaries of foreign firms as a growth model for latecomers. If indigenous firms can close the technology gap within a short span of time, then they can rapidly grow and catch up with foreign firms in developed countries. Then the arrow of technology acquisition of indigenous firms in Figure C.1 can be the same in length with that of foreign firms.

The second option is to select diversification as a corollary to compensate for the technology gap. This has been the focus of this book. Recall that the make-or-buy decision of Chinese electronics firms has been fulfilled precisely because of the development of the vertical specialization in the electronics industry, and the Chinese vast and diversified market, which allows indigenous firms to exert the home advantage. The conditions for external technology and internal knowledge will vary by time, country, industry, and firm. Therefore, the direction of diversification will also vary depending on the conditions faced by indigenous firms. The organizations of indigenous firms depend on the lengths of external technology, the technology acquisition, and internal knowledge, and the contents of external technology and internal knowledge.

In some cases, indigenous firms cannot enter and grow because the technology gap is despairingly large and cannot be compensated by both external technology and internal knowledge. This is the third option. Since indigenous firms cannot exist in markets, therefore it turns out that the boundaries of indigenous firms do not exist, that is, zero boundaries. As shown in Young (1991), industrial development in developing countries would be inhibited by the existence of competitive foreign firms in developed countries.

Thus, by focusing on the balance of the three factors, the homogenized and zero boundaries can be placed at either end of diversification. Although each of the homogenized, heterogenized, and zero boundaries have been already studied individually, as shown in the literature review in Chapter 1, they can be integrated in the present study's diversification mechanism. Consequently, it shows that globalization can provide an impetus to the rich diversity of the organizations of firms, although it possibly provides the convergence of technological levels between indigenous and foreign firms and possibly inhibits the entry and growth of indigenous firms.

C.2 Interactions between Firms with Different Advantages

The diversification mechanism can become a development mechanism for the whole of firms under competition or for an entire industry, particularly when the interaction itself between firms is focused on. Although this study has exclusively focused on the growth process of indigenous firms, the leading players highlighted in this book, emphasis can be placed on the interaction between firms with different advantages. To do so, we must reconsider the meaning of competition from the viewpoint of our diversification mechanism.

First, the technology gap is reconsidered as a precondition for diversification. It comprises two areas of indigenous firms' technology in comparison to that of foreign firms, unless indigenous firms do not acquire any technologies. First is an area in which indigenous firms can acquire technologies through technology diffusion, accumulation, and development. In this area, technologies owned by them overlap one another. The remainder of the overlapping area is the technology gap. Indigenous firms can use the advantage of backwardness; however, they probably face another area in which they cannot completely acquire technologies. This gap area in which indigenous firms cannot acquire technologies is also included such that they themselves decide not to learn technologies from foreign firms or develop technologies by themselves precisely because acquisition is irrational for economic reasons. If technologies are too advanced, then it would be irrational for indigenous firms to try to acquire them. Since there is a gap that indigenous firms cannot makeup, we have investigated the diversification of indigenous firms through competition with foreign firms.

Regarding the influence of competition on the whole of firms as opposed to the growth of individual firms, the area of technology acquisition, rather than that of technology gap, must first be focused upon. The former area can expand when indigenous firms make efforts to acquire advanced and sophisticated technologies. As a result, the average of the technological level for both indigenous and foreign firms can elevate and become the new standard that is referred to by firms when they develop new models to be differentiated from present ones. Therefore, the technology advantage can be relatively weakened because indigenous firms acquire the exclusive advantage of foreign firms. Consequently, foreign firms followed by indigenous firms need to dig deeper into the technology level. This can create the momentum to improve the technology advantage of foreign firms as we can extract this implication from the product lifecycle (PLC) model developed by Krugman (1979).

Next, another gap between the home advantage of indigenous firms and the away disadvantage of foreign firms has been focused on. As shown in this book, this is a gap that is excavated and developed by indigenous firms in order to offset the technology gap created by the difference in timing of the entries of indigenous and foreign firms. When the efficiency of activities based on certain advantages of competitors is revealed through competition, firms learn the competitive strategies or rather adopt them to get ahead in the competition. As discussed in Chapter 4, major foreign handset manufacturers caught up with the marketing-oriented strategy of Chinese firms and organized part of the distribution channels. The fact that foreign handset manufacturers learned the marketing-oriented strategy shows that Chinese handset firms led a new stage of industrial development by adding the latest standard in marketing on the list of capabilities of which the efficiency was already verified. Consequently, we can say that both indigenous and foreign firms have been developing China's handset industry through fierce competition.

In this way we have focused our attention on the competition only between indigenous and foreign firms in this book; however, our discussion can also be applied to general competition because any firm in a market can face a technology gap among firms in the market. Competition is usually developed among heterogeneous firms, not among homogeneous ones. In reality, it is exceptionally difficult to assume that all of the players in a market have equal levels of competitiveness and technology. This book has focused on just a combination of firms, that is, indigenous firms in developing countries and foreign firms in developed countries, but this is a significant combination to consider a development problem in globalization.

Therefore, we can apply our discussion to the effect of the interaction between firms with different technological levels and different advantages on the whole development in an industry. Every firm in a market absorbs other firms' technologies or advantages wherever possible and, at the same time, exerts own technologies or advantage to make up for gaps that the firm could not acquire other firms' technologies or advantages. Consequently, firms generally conduct the behavior that they absorb others' advantages as they can and differentiate own advantage to compensate for gaps and to differentiate competitiveness. Of course such a function of competition has been partially discussed (Numagami et al., 1992). However, by analyzing the balance among the three factors, we have revealed the optimal behavior from the viewpoint of one of the players in competition. Based on this discussion, we intend to determine the effect of competition among the whole players.

As Chinese firms have influenced the standard of capabilities in the Chinese handset market, are Chinese firms presently producing global knowledge at present in the electronics industry? Since Chinese global firms, such as Lenovo, Haier Group (Haier), Huawei Technologies (Huawei), ZTE, and so on, have emerged partially on the basis of capabilities accumulated through fierce competition in the domestic market, therefore they are receiving global attention. However, generally speaking, the overseas expansion of Chinese electronics firms has just begun. Most of them still have just small market shares for their electronics products in various countries, with some exceptions. Since the advantage of Chinese firms is based on the home market, therefore it is difficult for indigenous firms to exert it in foreign countries. Since our study has been focusing on the relative long-term growth process of indigenous firms, therefore recent changes have been excluded. Major Chinese firms have also begun to focus on R&D activities and continuously acquire technologies. Moreover, with foreign firms accumulating knowledge for the Chinese market through business operations and the development of the distribution industry, the home advantage has been undermined. In the near future, more Chinese firms will strengthen their presence in the global market by exerting capabilities accumulated through competition in the Chinese market. Chinese firms, which seized the opportunity to grow in the domestic market by using the inherent home advantage, will seize other opportunities to grow in the global market, using another's acquired home advantage. What strategies will foreign firms need to learn from them when Chinese firms dominate the global market? The global economy can develop through firms characterized by fierce competition among firms with different advantages.

Notes

Introduction

1. However, we specifically define the indigenous and foreign firms as necessary.
2. Indigenous firms are also known as domestic firms.
3. Discussions at the country level are not differenciated from those at the firm level, because, in this book, problems do not occur without a distinction between the macro and micro levels.
4. In addition to technology diffusion, the influence of globalization on the economic growth in developing countries depends on terms of trade, mobility of capital, and so on (Acemoglu, 2009; Bardhan and Udry, 1999).
5. This is contrary to vertical integration, that is, a firm integrates the production stages of upstream and/or downstream stages in value chains. Depending on the context, vertical specialization is also called the intra-industry division of labor, the vertical disintegration, and the horizontal specialization.

Chapter 1

1. One of other questions is about the channels of technology diffusion.
2. Studies on the influence of openness include a variety of viewpoints, such as the relationship between openness and the economic growth, not only in developing countries but also in developed countries; the relationship between large and small countries; and so on. In addition to the influence of globalization on growth, the relationship between globalization and growth also includes the influence of growth on the global economy (Feenstra, 2004).

3. Even if countries can access the technology stocks of trade partners through openness, there possibly exists a possible disadvantage that competition in technology development can heat up among them (Eaton and Kortum, 2001). Therefore, this disadvantage can diminish the positive effect of openness.
4. See Fukagawa (1989) for the development process of each industry and technology transfer from developed countries to Korea.
5. In addition, Keller considered geographical proximity as a factor of technology diffusion.
6. Related to the ability, Borensztein et al. (1998) showed that inward FDI has a positive effect when human capital is measured by education attainment (secondary schooling).
7. Other policies also can mitigate the negative effect. For example, Bhagwati (2004), who has stated that globalization fundamentally promotes the growth in developing countries, has emphasized the necessity of certain measures for the class suffering losses due to openness.
8. Note that the report was written not only to promote trade protection but also to initiate the importance of manufacturing in comparison to agriculture to further economic growth in the US.
9. To focus on a comparison between the cost and benefit of industrial development, Mill–Bastable has been used as the criteria of protection, even after Kemp's studies. For instance, Melitz (2005) uses the Mill–Bastable criteria to analyze optimal trade policies.
10. In contrast, Yasheng Huang (2003) indicated that the huge inflow of FDI shows the weakness of the Chinese economy. According to him, FDI has been flowing in to fill the weakness in the private sector and fragmented markets in China.
11. However, the influence of inward FDI on industrial development depends on local governments. Thun (2006) compared the automotive industries in Shanghai, where the local government exercises a strong leadership; in Beijing and Guangzhou, where the local governments take a relative laissez-faire approach; and in Changchun and Wuhan, where indigenous firms exert a strong leadership.
12. At present, as wages are increasing in China, the cost advantage is decreasing remarkably. In addition, manufacturers have been

facing numerous challenges, such as product differentiation through innovation, responding to environmental regulations, stronger yuan, and so on (Hu and Jefferson, 2008; Kimura, 2008, 2010a).
13. Many studies have been conducted on the industrial development in China's electronics industry. See Amano (2005), *Dangdai Zhongguo* Congshu Bianji Bu (1987), Fan (2004), Kimura (2007a), Kong (2006), Marukawa (1996), Nakagawa (2007), Ning (2009), Pecht et al. (1999), Popkin and Iyengar (2007), Tang (2009), Xie (2000), Zhao (2000), *Zhongguo Dianzi Gongye 50 nian* Bianweihui (1999), and so on.
14. Since industrial agglomerations have developed in China, especially on the coastal area, it has become easier to purchase intermediate goods and services for firms located in China (Kato, 2003; Kimura, 2007b).
15. Note that modularization possibly weakens the incentive for innovation, even though it has undoubtedly contributed to industrial development (Nakagawa, 2007).
16. In addition, since the problem of the default risk of trade credit was increasing in the 1990s, there was also the case of TCL wherein some manufacturers integrated part of distribution functions in order to avoid a default risk (Watanabe, 2010).
17. In addition to the property rights approach (property rights theory: PRT) as in the GHM model, there is also the transaction cost approach (transaction cost economics: TCE). According to TCE, by comparing costs between transactions inside firms and through markets, the boundaries are decided in order to minimize the costs. A major difference between these two approaches is that PRT places weight on holding down opportunistic behavior at the stage of enforcement after contracting, while TCE does this by holding down before contracting (Furubotn and Richter, 2005).
18. The theory of the boundaries of the firm has originated from studies on the existential reason of a firm. The reason was verified by Coase (1937) based on the concept of the transaction cost. The transaction cost is caused by transactions, such as search for information, bargaining between buyers and sellers, and enforcement of contracts. Williamson (1985) developed the idea of the cost and contributed to developing the TCE. Although the TCE clarifies the transaction cost in markets, it

Chapter 2

1. They are Chapters 84 and 85 of the HS code, respectively. Note that the chapters include some other products usually not to be included as electrical and electronics products.
2. Structures of sales networks have varied from one indigenous firm to another. See Zheng (2012) for similarities and differences in their sales activities.
3. Note that this table includes just the input coefficients of intra-industry industries, although input coefficients tables usually include all of inter-industry transactions among all combinations.

Chapter 3

1. A revised version of Kimura (2012a) is cited in this chapter.
2. The TFP growth rate itself is high in China, although it has been slowing down (Islam et al., 2006).
3. Among 8,391 firms, 3,287 manufacturers are JVs.
4. It may seem that a ratio of sales is more adequate as a variable of market share for capturing the market-stealing effect. However, since our purpose is to verify the relationship between the growth of indigenous and the investment by foreign capital, we use the ratio of fixed assets.
5. According to the traditional regional classification after the Seventh Five-Year Plan (1986–1990) and the level of development, a classification is set as follows in this chapter. The eastern region comprises Beijing, Tianjin, Hebei, Liaoning, Shandong, Shanghai, Jiangsu, Zhejiang, Fujian, Guangdong, and Hainan. The central region comprises Shanxi, Jilin, Heilongjiang, Anhui, Jiangxi, Henan, Hubei, Hunan, Inner Mongolia, and Guangxi. The western region comprises Sichuan, Chongqing,

---

does not reveal why internal transaction within the firm can avoid the transaction cost. Therefore, Grossman, Hart, and Moore assume that owners of the property right of a firm can control everything that does not draw unforeseen contingencies, thereby solving the transaction cost problem. This is, therefore, called the PRT.

Guizhou, Yunnan, Tibet, Shaanxi, Gansu, Qinghai, Ningxia, and Xinjiang.
6. Previous studies also indicate differences in productivity and its growth by ownership. Chen et al. (1988) and Jefferson et al. (1996) showed that TFP growth rates vary by ownership. In addition, Brambilla (2006) showed that foreign firms located in China are more productive than indigenous firms by 18 percent, using data for 1,500 firms from the *Investment Climate Survey* (the 2001 edition) published by the World Bank.
7. In addition, we do not have information about ages of firms or dates of establishment for every firm.

CHAPTER 4

1. A revised version of Kimura (2011b) is cited in this chapter.
2. In addition, they are also significantly expanding businesses in the global market. They are a top global manufacture at present.
3. Only 10 percent subscribed to the the CDMA system, and the use of the TD-SCDMA system had just begun as a test operation in April 2008. In addition to these systems, there are two telecommunication carriers operating in China: China Mobile and China Unicom. China Mobile operated the GSM and TD-SCDMA systems, while China Unicom used GSM and CDMA at the time. In 2008, the telecommunication industry was reorganized by the government. After the reorganization, the 3G system started full-scale introduction in 2009. Immediately after, there was a boom in the popularity of smartphones. Therefore, the handsets industry began gradually transitioning to the next stage.
4. The eight researches were conducted with Professor Zhijia Yuan, the late Mr. Kenichi Imai, Dr. Momoko Kawakami, Professor Tomoo Marukawa, Mr. Hideto Akiba, Professor Masanori Yasumoto, Associate Professor Hirohumi Tatsumoto, Mr. Jing-Ming Shiu (their affiliations are mentioned in the Acknowledgments), although members of each research have been different: (1) Beijing, Tianjin, Hangzhou, Suzhou, Ningbo, Shanghai, Nanjing, Huizhou, Guangzhou, and Shenzhen (August 22–September 10, 2004); (2) Beijing and Shanghai (September 25–October 1, 2005); (3) Guangzhou

and Shenzhen (November 1–5, 2005); (4) Shenzhen, Xiamen, Shanghai, and Guangzhou (July 25 to August 5, 2007); (5) Shanghai (January 23–27, 2007); (6) Xiamen (April 23–25, 2007); (7) Shanghai (June 6–8, 2007); and (8) Beijing and Shanghai (August 22–29, 2007). Moreover, the author conducted frequent interviews at related firms and organizations in Beijing when the author lived in Beijing from June 2005 to June 2007.
5. The mobile phone service was initiated in Guangdong in 1987.
6. In 2001, they established a joint venture (JV) with Sony (Japan), calling it Sony Ericsson (the United Kingdom). Finally, in 2011, Sony acquired Ericsson's stake, making the handset business a wholly subsidiary.
7. Presently known as Panasonic.
8. However, the data does not necessarily depict the real picture of the market. First, the data indicates shipping volume from manufacturers, not market share at the retail level. Therefore, the data can include channel inventories at the distribution level. Second, since the data is "official" ones, therefore they do not include shipping volumes of illegal handsets. See footnote number 11 of this chapter for the definition of illegality.
9. At the time, since China had not joined the World Trade Organization (WTO); thus, adopting such a favorable policy was not as restricted as it is at present.
10. Note that other indigenous firms that could not earn the license were also not allowed to enter the market.
11. Handsets were considered illegal if they (1) were manufactured by firms without a license, (2) did not have the requisite certification to connect to the mobile phone networks, (3) were smuggled, and (4) violated intellectual property rights. However, as mentioned, the licensing system was cancelled. Therefore, the type (1) illegality was no longer applicable, but the production of type (4) continues to increase with the rapidly growing market.
12. Since we focus on the relationship between firms with different technological levels, we include Taiwanese firms as foreign firms because of its higher technological capabilities. Taiwan's consumer electronics industry started growing rapidly since the late 1960s, while China's industry has started development since the 1980s (Sato, 2007). In *China Statistics*

Yearbook published by the Chinese government, although "enterprises with funds from Hong Kong, Macao and Taiwan" are not included in "domestic funded enterprises," though also not included in "foreign funded enterprises." It is preferable to consider the differences between Chinese and Taiwanese firms.
13. It was not that the whole elements of the OS and the middleware were modularized, so the bold line contains just part of the elements.
14. In addition, they launched major advertising campaigns, in particular via the television, to promote recognition of their new brand names.
15. The actual distribution stages are more complicated and longer, when firms sell products in rural areas.
16. This channel was formed in the late 1990s. Before that, telecommunication carriers had sold handsets by themselves (Ouyang, 2006).
17. For a discussion on some of major roles played by chip vendors in industrial development, see Shiu and Imai (2007, 2010).

Chapter 5

1. A revised version of Kimura (2011c) is cited in this chapter.
2. To this effect, it is assumed that there is no difference in entry between the modes of exports and outward FDI.
3. Other components and materials that make up the final goods are not of interest to the present study.
4. We do not consider the case of $z = 1/2$, because the optimal boundary cannot be decided in making or buying when $z = 1/2$.
5. The optimal amount is x_h when differentiating the profit function for the South firm with respect to the headquarters service (x_h). Similarly, the optimal amount is x_t when differentiating the profit function of the supplier with respect to the technology service (x_t). After substituting both the optimal amounts in the derivatives of the South firm and the supplier, respectively, the final optimal amounts are x_h and x_t. Finally, substituting the final optimal amounts in the price function, $p = \lambda^{1-\alpha}\sigma^{\alpha-1}x_h^{(\alpha-1)(1-z)}x_t^{(\alpha-1)z}$, which can be derived from Equations (5.1) and (5.2), results in Equation (5.7).

6. If a Cournot model of competition is assumed, then an optimal boundary selection is established on basis of the differences in optimal production volume, which depends on the technological level of the South firm. However, this chapter focuses on the entry conditions for the South firm, by comparing the cost levels of the North and South firms. The Bertrand competition model can compare the prices of both the firms.
7. As described above, $1/(1 - \alpha)$ is the price elasticity of demand $[-(dy/dp)/(y/p)]$ and the constant in our model. In the case of monopoly, the marginal revenue (MR) is as follows: $MR = (dp/dy)y + p = p[(dp)/(dy)/(p/y)+1] = p\{1 - 1/[-(dy/dp)/(y/p)]\} = \alpha p$. Therefore, multiplying the monopoly price by α results in MC.

Chapter 6

1. FDI is defined as investment that has a controlling interest in an investment-grade firm, which in essence means acquiring more than 10 percent of voting common stock.
2. The objectives and entry modes of outward FDI of indigenous Chinese firms have been investigated by Amano and Oki (2007), Buckley et al. (2007), Jiang (2011), Kawai (2013), Lu and Yan (2011), Marukawa and Nakagawa (2008), Ohashi (2003), Takahashi (2008), Xu (2010), and Zhang (2009), among others. Deng (2012) provided a detailed literature review on outward FDI from China.
3. Dunning and Lundan (2008) uses the OLI framework to elucidate the determinants of outward FDI. "OLI" combines the initials of "ownership, location, and internalization" advantages. Having the ownership advantage means that investors must have competitiveness in terms of technologies, management know-how, and brand names. The location advantage implies that host countries must have advantages such as low wages, natural resources, and high tariff rates. With the internalization advantage, investors possess rationality to invest due to tacit knowledge and/or trade secret(s). In this study, the ownership advantage is defined as the competitiveness of investors. Helpman et al. (2004) and Antràs and Helpman (2004) showed that productive firms tend to invest

and export. Their studies focused on heterogeneity among firms and showed that the most productive firms are known to be investors, followed by exporters and firms focusing on domestic markets.
4. A significant number of empirical studies show that the productivity of investors is higher than that of noninvestors.
5. In addition to examining the characteristics of MNEs from developing countries, Kawai (2013) has discussed the characteristics of Chinese MNEs, which have played a positive role in attracting overseas investments in Southeast Asian countries. There are various factors influencing the investment decision of Chinese firms.
6. The five field researches are as follows: (1) Beijing (June 22–26, 2010); (2) Johannesburg (September 4–12, 2010) with Dr. Takahiro Fukunishi (IDE); (3) Shenzhen (November 30 to December 4, 2010); (4) Johannesburg (September 3–11, 2011); and (5) Qingdao (October 27–29, 2011).
7. Since Chinese firms are foreign firms from the viewpoint from SA, in this chapter, "indigenous firms" refer to firms in SA.
8. The SA market has continued to grow in the 2000s with the boom in natural resources. The country is included among the fast-growing countries known as "BRICS."
9. Gelb (2010) and Kimura (2012b, 2012c, 2013b) researched the activities of Chinese home appliance and consumer electronics manufacturers in SA. On the other hand, this study focuses on the relationships with indigenous SA firms as a determinant or a condition for outward FDI.
10. The "prima" in Sinoprima stands for the company's brand name used in markets other than in SA. In the SA market it is known as "Sinotec."
11. Internationalization strategies of Chinese MNEs depend on firms. Although Haier Group (Haier) tends to enter developed markets before entering emerging markets, TCL tends to enter emerging markets before entering developed markets (Kojima, 2007).
12. TV sets that are not categorized as "ordinary" are exempted from import tariffs. These TV sets with screens that do not exceed 45 centimeters, or those with a screen size exceeding 3 meter × 4 meter. All black-and-white TV sets are made using CRTs.

13. By comparison, complete-knockdown (CKD) production is defined as the assembling of various components and PCBs that are not mounted semiconductor parts. Therefore, firms must mount semiconductor parts on the surface of PCBs (interview at Sony in Johannesburg, on September 8, 2011).
14. An issue faced by Chinese home appliance manufacturers in SA is the loss of cost competitiveness when expanding local production. As far as Chinese MNEs perform KD production using duty-free components, they are able to retain their price competitiveness. This issue also applies to Chinese manufacturers engaged in local production in other foreign countries. In other words, their cost competitiveness cannot be separated from the low wages in China.
15. Store brand is also known as private label, private brand, house brand, and so on.
16. There has been a case where a Chinese firm acquired by a European company entered the SA market under a European brand name (Kimura, 2012c). Thus although Chinese firms generally do not have strong brand names, there are various ways to compensate for this problem and to enter overseas markets.

References

Abe, Makoto (2006) "Kankoku Keitai Denwa Tanmatsu Sangyo no Seicho: Denshi Sangyo tono Renzokusei to Hirenzokusei kara [Growth of Korea's Mobile Handset Industry: Its Continuity and Discontinuity with the Development of the Electronics Industry]," in Kenichi Imai and Momoko Kawakami, eds., *Higashi Azia no IT Kiki Sangyo: Bungyo, Kyoso, Sumiwke no Dainamikusu* [*The Information Technology Equipment Industry in East Asia*], Chiba: Institute of Developing Economies: 17–53 (in Japanese).

Acemoglu, Daron (2009) *Introduction to Modern Economic Growth*, Princeton: Princeton University Press.

Aghion, Philippe and Peter Howitt (2009) *The Economics of Growth*, Cambridge: The MIT Press.

Aghion, Philippe and Rachel Griffith (2005) *Competition and Growth: Reconciling Theory and Evidence*, Cambridge: The MIT Press.

Aitken, Brian and Ann Harrison (1999) "Do Domestic Firms Benefit from Direct Foreign Investment? Evidence from Venezuela," *American Economic Review* 89(3): 605–618.

Amano, Tomofumi (2005) "Chugoku Kaden Sangyo no Hatten to Nihon Kigyo: Nicchu Kaden Kigyo no Kokusai Bungyo no Tenkai [The Development of Chinese Electric Industry and Japanese Firms: The Transition of International Division of Labor]," *Journal of JBIC Institute* No. 22: 116–134 (in Japanese).

Amano, Tomofumi and Hiromi Oki, eds. (2007) *Chugoku Kigyo no Kokusaika Senryaku* [*Internationalization Strategies of Chinese Firms*], Tokyo: Japan External Trade Organization (in Japanese).

Amsden, Alice H. (1989) *Asia's Next Giant: South Korea and Late Industrialization*, New York: Oxford University Press.

Ando, Mitsuyo and Fukunari Kimura (2005) "The Formation of International Production and Distribution Networks in East

Asia," in Takatoshi Ito and Andrew K. Rose, eds., *International Trade in East Asia*, Chicago: The University of Chicago Press: 177–216.

Antràs, Pol (2005) "Incomplete Contracts and the Product Cycle," *American Economic Review* 95(4): 1054–1073.

Antràs, Pol and Elhanan Helpman (2004) "Global Sourcing," *Journal of Political Economy* 112(3): 552–580.

Arai, Toru (2005) "Chugoku Kaden Ryutsu no Hensen to Tenbo [Transition and Prospect of Home Appliance Distribution in China]," in Hiroshi Matsue, ed., *Gendai Chugoku no Ryutsu [Distribution in Modern China]*, Tokyo: Dobunkan Shuppan: 187–202 (in Japanese).

Baldwin, Carliss Y. and Kim B. Clark (2000) *Design Rules, Volume 1: The Power of Modularity*, Cambridge: The MIT Press.

Baldwin, Robert E. (1969) "The Case against Infant-Industry Tariff Protection," *Journal of Political Economy* 77(3): 295–305.

Bardhan, Pranab and Christopher Udry (1999) *Development Microeconomics*, Oxford: Oxford University Press.

Barro, Robert J. and Xavier Sala-i-Martin (2004) *Economic Growth*, 2nd edn., Cambridge: The MIT Press.

Besanko, David, David Dranove, Mark Shanley and Scott Schaefer (2009) *Economics of Strategy*, 5th edn., Hoboken: John Wiley and Sons.

Bhagwati, Jagdish (2004) *In Defense of Globalization*, New York: Oxford University Press.

Björkstén, Johan and Anders Hägglund (2010) *How to Manage a Successful Business in China*, Singapore: World Scientific Publishing.

Borensztein, Eduardo, Jose De Gregorio and Jong-Wha Lee (1998) "How Does Foreign Direct Investment Affect Economic Growth?" *Journal of International Economics* 45(1): 115–135.

Brambilla, Irene (2006) "Multinationals, Technology and the Introduction of Varieties," *NBER Working Paper*, No. 12217.

Branstetter, Lee and Nicholas R. Lardy (2008) "China's Embrace of Globalization," in Loren Brandt and Thomas Rawski, eds., *China's Great Economic Transformation*, New York: Cambridge University Press: 633–682.

Breznitz, Dan and Michael Murphree (2011) *Run of the Red Queen: Government, Innovation, Globalization, and Economic Growth in China*, New Haven: Yale University Press.

Buckley, Peter J., L. Jeremy Clegg, Adam R. Cross, Xin Liu, Hinrich Voss and Ping Zhang (2007) "The Determinants of Chinese Outward Foreign Direct Investment," *Journal of International Business Studies* 38(4): 499–518.

Chen, Kuan, Hongchang Wang, Yuxin Zheng, Gary H. Jefferson and Thomas G. Rawski (1988) "Productivity Change in Chinese Industry: 1953–1985," *Journal of Comparative Economics* 12(4): 570–591.

Chen, Yifei (1994) *Kaituo Nanfei Shichang* [*Pioneering South African Market*], Beijing: China Social Sciences Press (in Chinese).

Coase, Ronald (1937) "The Nature of the Firm," *Economica* 4(16): 386–405.

Corden, W. Max (1997) *Trade Policy and Economic Welfare*, 2nd edn., Oxford: Clarendon Press.

Dangdai Zhongguo Congshu Bianji Bu (1987) *Danddai Zhongguo de Dianzi Gongye* [*Modern China's Electronics Industry*], Beijing: China Social Sciences Press (in Chinese).

Deng, Ping (2012) "The Internationalization of Chinese Firms: A Critical Review and Future Research," *International Journal of Management Review* 14(4): 408–429.

Djankov, Simeon and Bernard Hoekman (2001) "Foreign Investment and Productivity Growth in Czech Enterprises," *World Bank Economic Review* 14(1): 49–64.

Dong, Zhikai and Jiang Wu (2004) *Xin Zhongguo Gongye de Dianjishi: 156 xiang Jianshe Yanjiu* [*Industry Cornerstone of New China: A Study on 156 Constructions*], Shenzhen: Guangdong Economy Publishing House (in Chinese).

Dunning, John and Sarianna M. Lundan (2008) *Multinational Enterprises and the Global Economy*, 2nd edn., Cheltenham: Edward Elgar.

Eaton, Jonathan and Samuel Kortum (2001) "Technology, Trade, and Growth: A Unified Framework," *European Economic Review* 45(4–6): 742–755.

Economic Structure Reform and Economic Operation Office of Ministry of Information Industry of China (2003) *Fazhanzhong de Woguo Shouji Chanye* [*Our Developing Mobile-Phone Handset Industry*], Beijing: Publishing House of Electronics Industry (in Chinese).

Fan, Jianting (2004) *Chugoku no Sangyo Hatten to Kokusai Bungyo: Taichu Toshi to Gijutsu Iten no Kensyo* [*China's Industrial*

Development and International Division of Labor: A Verification of Investment in China and Technology Transfer], Tokyo: Fukosha (in Japanese).

Fan, Yongming (1992) *Zhongguo de Gongyehua yu Waiguo Zhijie Touzi* [*China's Industrialization and Foreign Direct Investment*], Shanghai: Shanghai Academy of Social Sciences Press (in Chinese).

Feenstra, Robert C. (2004) *Advanced International Trade: Theory and Evidence*, Princeton: Princeton University Press.

Fujimoto, Takahiro and Junjiro Shintaku, eds. (2005) *Chugoku Seizogyo no Akitekucha Bunseki* [*Architecture-based Analysis of China's Manufacturing Industries*], Tokyo: Toyo Keizai (in Japanese).

Fukagawa, Yukiko (1989) *Kankoku: Aru Sangyo Hatten no Kiseki* [*South Korea: A Trajectory of Industrial Development*], Tokyo: Japan External Trade Organization (in Japanese).

Fukagawa, Yukiko (1997) *Kankoku Senshinkoku Keizai Ron: Seijuku Katei no Mikuro Bunseki* [*South Korea—Developed Country Economics: A Micro Analysis on Maturing Process*], Tokyo: Nikkei (in Japanese).

Furubotn, Eirik G. and Rudolf Richter (2005) *Institutions and Economic Theory: The Contribution of the New Institutional Economics*, Ann Arbor: The University of Michigan Press.

Gelb, Stephen (2010) "Foreign Direct Investment Links between South Africa & China," *The EDGE Institute Discussion Paper*.

Gerschenkron, Alexander (1962) "Economic Backwardness in Historical Perspective," in Alexander Gerschenkron, ed., *Economic Backwardness in Historical Perspective: A Book of Essays*, Cambridge: Belknap Press of Harvard University Press: 5–30.

Girma, Sourafel (2005) "Absorptive Capacity and Productivity Spillovers from FDI: A Threshold Regression Analysis," *Oxford Bulletin of Economics and Statistics* 67(3): 281–306.

Grossman, Gene M. and Elhanan Helpman (1991) *Innovation and Growth in the Global Economy*, Cambridge: The MIT Press.

Grossman, Sanford and Oliver Hart (1986) "The Costs and Benefits of Ownership: A Theory of Vertical and Lateral Integration," *Journal of Political Economy* 94(4): 691–719.

Hamilton, Alexander (1791) *The Report on the Subject of Manufactures*. Reprinted in Harold C. Syrett, ed. (1966) *The Papers of*

Alexander Hamilton, Vol. 10, New York and London: Columbia University Press.
Hao, Yanshu (1999) *Chugoku no Keizai Hatten to Nihon teki Seisan Shisutemu* [*China's Economic Development and Japanese Production System*], Kyoto: Minerva Shobo (in Japanese).
Harrison, Ann and Ana Revenga (1995) "The Effects of Trade Policy Reform: What Do We Really Know?" *NBER Working Paper*, No. 5225.
Harrison, Ann and Andres Rodríguez-Clare (2010) "Trade, Foreign Investment, and Industrial Policy for Developing Countries," *Handbook of Development Economics, Vol. 5*, New York: Elsevier.
Hart, Oliver and John Moore (1990) "Property Rights and the Nature of the Firm," *Journal of Political Economy* 98(6): 1119–1158.
Helpman, Elhanan, Dalia Marin and Thierry Verdier, eds. (2008) *The Organization of Firms in a Global Economy*, Cambridge: Harvard University Press.
Helpman, Elhanan, Marc J. Melitz and Stephen R. Yeaple (2004) "Export Versus FDI with Heterogeneous Firms," *American Economic Review* 94(1): 300–316.
Hiratsuka, Daisuke and Fukunari Kimura, eds. (2008) *East Asia's Economic Integration: Progress and Benefit*, New York: Palgrave Macmillan.
Hobday, Michael (1995) *Innovation in East Asia: The Challenge to Japan*, Cheltenham: Edward Elgar.
Hu, Albert G. Z. and Gary H. Jefferson (2002) "FDI Impact and Spillover: Evidence from China's Electronic and Textile Industries," *The World Economy* 25(8): 1063–1076.
Hu, Albert G. Z. and Gary H. Jefferson (2008) "Science and Technology in China," in Loren Brandt and Thomas Rawski, eds., *China's Great Economic Transformation*, New York: Cambridge University Press: 286–336.
Huang, Lin (2003) *Shinko Shijo Senryaku Ron: Gurobaru Nettowaku to Maketingu Inobeshon* [*Strategies in Emerging Markets: Global Networks and Marketing Innovation*], Tokyo: Chikura Publishing (in Japanese).
Huang, Yasheng (2003) *Selling China: Foreign Direct Investment during the Reform Era*, New York: Cambridge University Press.
Imai, Ken and Jing Ming Shiu (2011) "Value Chain Creation and Reorganization in China," in Momoko Kawakami and Timothy

J. Sturgeon, eds., *The Dynamics of Local Learning in Global Value Chinas: Experiences from East Asia*, Basingstoke and New York: Palgrave Macmillan: 43–67.

Imai, Kenichi and Jingming Shiu (2008) "Shijo Kibo to Sangyo Kodoka: Keitai Denwa Sangyo no Kesu [Market Size and Industrial Upgrading: A Case of Mobile Industry]," in Kenichi Imai and Ke Ding, eds., *Chugoku: Sangyo Kodoka no Choryu [China: A Trend of Industrial Upgrading]*, Chiba: Institute of Developing Economies: 13–45 (in Japanese).

Irwin, Douglas A. (1996) *Against the Tide: An Intellectual History of Free Trade*, Princeton: Princeton University Press.

Ito, Banri, Naomitsu Yashiro, Zhaoyuan Xu, Xiaohong Chen and Ryuhei Wakasugi (2012) "How Do Chinese Industries Benefit from FDI Spillovers?" *China Economic Review* 23(2): 342–356.

Ito, Keiko, Moosup Jung, Young Gak Kim and Tangjun Yuan (2008) "A Comparative Analysis of Productivity Growth and Productivity Dispersion: Microeconomic Evidence Based on Listed Firms from Japan, Korea and China," *Seoul Journal of Economics* 21(1): 39–91.

Ito, Motoshige, Kazuharu Kiyono, Masahiro Okuno and Kotaro Suzumura (1988) *Sangyo Seisaku no Keizai Bunseki [Economic Analysis on Industrial Policy]*, Tokyo: University of Tokyo Press (in Japanese).

Islam, Nazrul, Erbiao Dai and Hiroshi Sakamoto (2006) "Role of TFP in China's Growth," *Asian Economic Journal* 20(2): 127–159.

Jefferson, Gary H., Thomas G. Rawski and Yuxin Zheng (1996) "Chinese Industrial Productivity: Trends, Measurement Issues and Recent Developments," *Journal of Comparative Economics* 23(2): 146–80.

Jensen, Bjarne S. and Kar-yiu Wong, eds. (1997) *Dynamics, Economic Growth, and International Trade*, Ann Arbor: The University of Michigan Press.

Jiang, Dianchun and Yu Zhang (2006) "Hangye Tezheng yu Waishang Zhijie Touzi de Jishu Yichu Xiaoying: Jiyu Gaoxin Jishu Chanye de Jingyan Fenxi [Industrial Characteristics and Technology Spillover of FDI: The Empirical Evidence of Chinese High-Tech Industries]," *Journal of World Economy*, 10: 21–29 (in Chinese).

Jiang, Hongxiang (2011) "Chugoku no 'Soushutsukyo' Seisaku to Taigai Chokusetsu Toshi no Sokushin: Gijutsu Kakutoku o Chushin ni [A Study of Foreign Direct Investment Promotion through Chinese 'Go Abroad' Policy: Focusing on Technology Acquisition]," *Ryukoku Journal of Economic Studies* 51(1): 21–49 (in Japanese).

Jiang, Xiaojuan (1993) *Zhongguo Gongye Fazhan yu Duiwai Jingji Maoyi Guanxi de Yanjiu* [*Industrial Development and Foreign Economic Relations in China*], Beijing: Economy & Management Publishing Press (in Chinese).

Johanson, Jan and Jan-Erik Vahlne (1977) "The Internationalization Process of the Firm: A Model of Knowledge Development and Increasing Foreign Market Commitments," *Journal of International Business Studies* 8(1): 23–32.

Johanson, Jan and Jan-Erik Vahlne (2009) "The Uppsala Internationalization Process Revisited: From Liability of Foreignness to Liability of Outsidership," *Journal of International Business Studies* 40(9): 1411–1431.

Joseph, K. J. (2004) "The Electronics Industry," in Subir Gokarn, Anindya Sen and Rajendra R. Vaidya, eds., *The Structure of Indian Industry*, New Delhi: Oxford University Press.

Kato, Hiroyuki (2003) *Chiiki no Hatten* [*Development of the Region*], Nagoya: University of Nagoya Press (in Japanese).

Kawai, Shinichi, ed. (2013) *Chugoku Takokuseki Kigyo no Kaigai Keiei* [*Overseas Operations of Chinese Multinational Enterprises*], Tokyo: Nippon Hyoron Sha (in Japanese).

Kawakami, Momoko (2012) *Asshuku sareta Sangyo Hatten* [*Compressed Industrial Development*], Nagoya: University of Nagoya Press (in Japanese).

Kawakami, Momoko and Timothy J. Sturgeon, eds. (2011) *The Dynamics of Local Learning in Global Value Chinas: Experiences from East Asia*, Basingstoke and New York: Palgrave Macmillan.

Kemp, Murray C. (1960) "The Mill-Bastable Infant-Industry Dogma," *Journal of Political Economy* 68(1): 65–67.

Kemp, Murray C. (1964) *The Pure Theory of International Trade*, Englewood Cliffs: Prentice-Hall.

Keller, Wolfgang (1996) "Absorptive Capacity: On the Creation and Acquisition of Technology in Development," *Journal of Development Economics* 49(1): 199–277.

Keller, Wolfgang (2004) "International Technology Diffusion," *Journal of Economic Literature* 42(3): 752–782.

Kim, Linsu (1997) *Imitation to Innovation: The Dynamics of Korea's Technological Learning*, Boston: Harvard Business School Press.

Kim, Linsu and Richard Nelson, eds. (2000) *Technology, Learning, and Innovation: Experiences of Newly Industrializing Economies*, Cambridge: Cambridge University Press.

Kimura, Fukunari and Kozo Kiyota (2006) "Exports, FDI, and Productivity: Dynamic Evidence from Japanese Firms," *Review of World Economics* 142(4): 695–719.

Kimura, Koichiro (2006a) "Macroeconomic Stability and Seigniorage for Fiscal Revenue: East Asia versus Eastern Europe and the CIS," in Mariko Watanabe, ed., *Recovering Financial Systems: China and Asian Transition Economies*, Basingstoke and New York: Palgrave Macmillan: 27–56.

Kimura, Koichiro (2006b) "Chugoku Keitai Denwa Tanmatsu Sangyo no Hatten: Hanbai Zhushi no Senryaku to Sono Genkai [Development of China's Mobile Industry: Marketing-Oriented Strategy and Its Limitations]," in Kenichi Imai and Momoko Kawakami, eds. *Higashi Azia no IT Kiki Sangyo: Bungyo, Kyoso, Sumiwke no Dainamikusu* [*The Information Technology Equipment Industry in East Asia*], Chiba: Institute of Developing Economies: 95–136 (in Japanese).

Kimura, Koichiro (2007a) "Yushutsu no Kenin Sangyo (2): Denki Denshi Sangyo [Leading Industries in Exporting (2): Electrical and Electronics Industry]," in Reiitsu Kojima and Nobuhiro Horii, eds., *Kyodaika suru Chugoku Keizai to Sekai* [*Growing Chinese Economy and the World*], Chiba: Institute of Developing Economies: 60–72 (in Japanese).

Kimura, Koichiro (2007b) "China's Regional Industrial Disparity from the Viewpoint of Industrial Agglomeration," in Masatsugu Tsuji, Emanuele Giovannetti and Mitsuhiro Kagami, eds., *Industrial Agglomeration and New Technologies: A Global Perspective*, Cheltenham: Edward Elgar: 173–201.

Kimura, Koichiro (2008) "Zai Chugoku Kigyo no RoHS Taio: Nikkei Kigyo to Genchi Kigyo [Addressing the RoHS Directive by Firms Located in China: Japanese and Indigenous Firms]," *World Eco Scope* (https://www.ecobrain-wes.com/wes/) posted on October (in Japanese).

Kimura, Koichiro (2010a) "Kokusai Kankyo Kisei to Denki Denshi Kiki Sangyo [International Environmental Regulations and Electrical and Electronic Equipment Industry]," in Nobuhiro Horii, ed., *Chugoku no Jizoku Kano na Seicyo: Shigen Kankyo Seiyaku no Kokufuku wa Kano ka?* [*China's Sustainable Growth: Whether Resource and Environmental Constraints Could Be Overcame or Not?*], Chiba: Institute of Developing Economies: 221–244 (in Japanese).

Kimura, Koichiro (2010b) "Chugoku no Keitai Denwa Tanmatsu Sangyo: Chugoku Ote Keitai Denwa Meka no Kyuseicho to Mosaku [China's Mobile Handset Industry: Rapid Growth and Challenges of Major Manufactures]," in Tomoo Marukawa and Masanori Yasumoto, eds., *Keitai Denwa Sangyo no Shinka Purosesu: Nihon ha Naze Koritsu Shita noka?* [*Evolution Process in the Mobile Industry: Why Japan has been Isolated?*], Tokyo: Yuhikaku: 173–195 (in Japanese).

Kimura, Koichiro (2011a) "China and India's Electrical and Electronics Industries: A Comparison between Market Structures," in Moriki Ohara, M. Vijayabaskar and Hong Lin, eds., *Industrial Dynamics in China and India: Firms, Clusters, and Different Growth Paths*, Basingstoke and New York: Palgrave Macmillan: 40–59.

Kimura, Koichiro (2011b) "Is There Hope for Firms Facing the Technology Gap? A Case of China's Mobile Industry," *Journal of Contemporary China* 20(72): 833–847.

Kimura, Koichiro (2011c) "Gijutsu Kakusa to Kigyo no Kyokai [The Technology Gap and the Boundaries of the Firm]," *Waseda Economic Studies* No. 70: 1–16 (in Japanese).

Kimura, Koichiro (2012a) "Does Foreign Direct Investment Affect the Growth of Local Firms? The Case of China's Electrical and Electronics Industry," *China & World Economy* 20(2): 98–120.

Kimura, Koichiro (2012b) "Genchi Brando no Moto deno Seicho: Minami Afurika no Chugoku Kigyo [Growth under Local Brand Names: Chinese Firms in South Africa]," *Ajiken World Trend* 18(11): 24–25 (in Japanese).

Kimura, Koichiro (2012c) "Gaishi to Ittai ni natta Minami Afurika Shinshutsu [The Entry in South Africa by Chinese Firms with Foreign Firms]," *Ajiken World Trend* 18(7): 29 (in Japanese).

Kimura, Koichiro (2013a) "Gijutsu: Kaigai kara no Gijutsu Donyu to Osei na Sannyu [Technology: Foreign Technology Introduction and Vigorous Entry]," in Mariko Watanabe, ed. *Chugoku no Sangyo wa Donoyoni Hatten Shitekitaka? [How Chinese Industries Have Developed?]*, Tokyo: Keiso Shobo: 212–233 (in Japanese).

Kimura, Koichiro (2013b) "Chugoku Kigyo no Minami Afurika Shinshutsu: Kaden Sangyo no Jirei [Chinese Firms in South Africa: The Case of the Home Appliance Industry]," in Kumiko Makino and Chizuko Sato, eds., *Minami Afurika no Keizai Shakai Henyo [Economic and Social Transformation in Democratic South Africa]*, Chiba: Institute of Developing Economies: 145–171 (in Japanese).

Kinoshita, Yuko (2001) "R&D and Technology Spillovers via FDI: Innovation and Absorptive Capacity," *CEPR Discussion Paper* No. 2775.

Kojima, Reiitsu (1997) *Gendai Chugoku no Keizai [Modern Chinese Economy]*, Tokyo: Iwanami Shoten (in Japanese).

Kojima, Reiitsu and Nobuhiro Horii, eds., (2007) *Kyodaika suru Chugoku Keizai to Sekai [Growing Chinese Economy and the World]*, Chiba: Institute of Developing Economies (in Japanese).

Kojima, Sueo (2007) "TCL [TCL]," in Tomofumi Amano and Hiromi Oki, eds., *Chugoku Kigyo no Kokusaika Senryaku [Internationalization Strategies of Chinese Firms]*, Tokyo: Japan External Trade Organization: 134–158 (in Japanese).

Kokko, Ari (1994) "Technology, Market Characteristics, and Spillovers," *Journal of Development Economics* 43(2): 279–293.

Komagata, Tetsuya (2000) "Gunji Kogyo: Gunmin Tenkan to sono Senryakuteki Haikei [Military Industry: Conversion and Its Strategic Background]," in Tomoo Marukawa, ed., *Ikoki Chugoku no Sangyo Seisaku [China's Industrial Policy in Transition]*, Chiba: Institute of Developing Economies: 293–334 (in Japanese).

Kong, Xinxin (2006) *Zhongguo Dianzi Gongye Chengzhang Dongli Yinsu Fenxi [Growth Dynamic Factors Analysis on China's Electronics Industry]*, Beijing: Economy & Management Publishing House (in Chinese).

Krugman, Paul R. (1979) "A Model of Innovation, Technology Transfer, and the World Distribution of Income," *Journal of Political Economy* 87(2): 253–266.

Krugman, Paul R., Maurice Obstfeld and Marc J. Melitz (2012) *International Economics: Theory & Policy*, 9th edn., Essex: Pearson Education.

Levinsohn, Jamese and Amil Petrin (2003) "Estimating Production Functions Using Inputs to Control for Unobservables," *Review of Economic Studies* 70(2): 317–341.

Li, Shaohua (2010) "Mokuaihua, Mokuai Zaizhenghe yu Chanye Geju de Chonggou: Yi 'Shanzhai' Shouji Jueqi Wei Li [Modularization, Modular Reintegration and Restruction of Industrial Structure: A Case of 'Shanzhai' Mobile's Rise]," *China Industrial Economy* 7: 136–145 (in Chinese).

Li, Xiaoxi (2009) "Duiwai Kaifang Lilun Yantao yu Jinzhan [Discussion and Progress of the Theory on the Open-Door Policy]," in Zhuoyuan Zhang, ed., *Zhongguo Jingjixue 60 nian: 1949– 2009 [60 Years of Studies on the Chinese Economy]*, Beijing: China Social Sciences Press: 469–515 (in Chinese).

Lin, Justin Yifu, Fang Cai and Zhou Li (1994) *Zhongguo de Qiji: Fazhan Zhanlüe yu Jingji Gaige [The China Miracle: Development Strategy and Economic Reform]*, Shanghai: Shanghai Joint Publishing Press and Shanghai People's Publishing House (in Chinese).

Ling, Zhijun (2005) *Lianxiang Fengyun [Lenovo's Storm]*, Beijing: China CITIC Press (in Chinese).

List, Friedrich (1841) *Das Nationale System der Politischen Oekonomie*, Jena: Gustav Fischer (in Germany). Translated as G. A. Mantile (1856) *National System of Political Economy*, Philadelphia: Lippincott.

Liu, Haiyan, ed. (2002) *Zhongguo Qiye shi: Dianxing Qiye juan [History of Chinese Firms: Volumes of Representative Firms]*, Vol. 1, Beijing: Enterprise Management Publishing House (in Chinese).

Long, Ngo Van and Kar-yiu Wong (1997) "Endogenous Growth and International Trade: A Survey," in Bjarne S. Jensen and Kar-yiu Wong, eds., *Dynamics, Economic Growth, and International Trade*, Ann Arbor: The University of Michigan Press.

Lu, Jinyong and Shiqiang Yan (2011) "Jingwai Zhijie Touzi Hangye Fenbu: Tedian, Yanbian he Qushi [Industrial Distribution of Chinese Outward Foreign Direct Investment: Characteristics, Evolution and Trend]," *Guoji Jingji Hezuo [International Economic Cooperation Journal]* 6: 22–26 (in Chinese).

Lu, Yue, Wei Xing and Hua Zhuan (2006) *Shouji Bantu: Muzhi Diguo de Jiaoliang* [*Mobile Empire: Tests of Strength among Thumb Empires*], Beijing: Guangming Daily Press (in Chinese).

Lucas, Robert E., Jr. (1988) "On the Mechanics of Development Planning," *Journal of Monetary Economics* 22(1): 3–42.

Mansfield, Edwin (1968) *The Economics of Technological Change*, New York: W. W. Norton.

Mansfield, Edwin, Anthony Romeo, Mark Schwartz, David Teece, Samuel Wagner and Peter Brach, eds. (1982) *Technology Transfer, Productivity, and Economic Policy*, New York: W. W. Norton.

Marukawa, Tomoo (1990) "Chugoku: Gijutsu Inten no Senryaku to Sisutemu [China: Technology Transfer Strategy and System]," in Takao Taniura, ed., *Azia no Kogyoka to Gijutsu Iten* [*Asia's Industrialization and Technology Transfer*], Tokyo: Institute of Developing Economies: 199–232 (in Japanese).

Marukawa, Tomoo (1996) "Shijo Keizai Iko no Purosesu: Chugoku Denshi Sangyo no Jirei kara [The Transition to Market Economy: In the Case of China's Electronics Industry]," *Ajia Keizai* 37(6): 2–28 (in Japanese).

Marukawa, Tomoo (2007) *Gendai Chugoku no Sangyo: Bokko Suru Chugoku Kigyo no Tsuyosa to Yowasa* [*Modern China's Industries: Strengths and Fragilities of Growing Chinese Firms*], Tokyo: Chuokoron-Shinsha (in Japanese).

Marukawa, Tomoo, Asei Ito and Yongqi Zhang (2014) *China's Outward Foreign Direct Investment Data*, Tokyo: Institute of Social Science, University of Tokyo.

Marukawa, Tomoo and Ryoji Nakagawa, eds. (2008) *Chugoku Hatsu Takokuseki Kigyo* [*Multinational Companies from China*], Tokyo: Doyukan (in Japanese).

Marukawa, Tomoo, Masanori Yasunoto, Kenichi Imai and Jingming Shiu (2006) "Nicchu Keitai Denwa Tanmatsu Sangyo no Hikaku [A Comparison between Mobile Handset Industries in Japan and China]," *Akamon Business Review* 5(8): 542–572 (in Japanese).

Maruyama, Nobuo (1988) *Chugoku no Kogyoka to Sangyo Gijutsu Shinpo* [*China's Industrialization and Industry Technology Advance*], Tokyo: Institute of Developing Economies (in Japanese).

Mayer, Thierry and Gianmarco Ottaviano (2008) "The Happy Few: The Internationalization of European Firms," Bruegel Blueprint Series Vol. 3, Brussels: Bruegel.

Meier, Gerald M. (1963) *International Trade and Development*, New York: Harper and Row.

Melitz, Marc J. (2003) "The Impact of Trade on Intra-Industry Reallocations and Aggregate Industry Productivity," *Econometrica* 71(6): 1695–1725.

Melitz, Marc J. (2005) "When and How Should Infant Industries be Protected?" *Journal of International Economics* 66(1): 177–196.

Ministry of Information Industry of China (2003) *Zhongguo Dianzi Gongye Nianjian (2003)* [*Yearbook of the Chinese Electronics Industry (2003)*], Beijing: Publishing House of Electronics Industry (in Chinese).

Ministry of Information Industry of China (2007) *2007 Zhongguo Xinxi Chanye Nianjian (Dianzi Juan)* [*Yearbook of the Chinese Information Industry (Volume of Electronics)*], Beijing: Publishing House of Electronics Industry (in Chinese).

Ministry of Information Industry of China (various years) *Zhongguo Dianzi Xinxi Chanye Tongji Nianjian* [*Statistical Yearbook of Chinese Electronics and Information Industry*], Beijing: Publishing House of Electronics Industry (in Chinese).

Mo, Bangfu (2013) *Sekai Shea Nanba Wan o Kakutoku shita Kokyaku Senryaku* [*Haier's Customer Strategies: How It Became the World's No. 1 Manufacturer of Major Appliances*], Tokyo: Chukei Shuppan (in Japanese)

Motohashi, Kazuyuki and Yuan Yuan (2010) "Productivity Impact of Technology Spillover from Multinationals to Local Firms: Comparing China's Automobile and Electronics Industries," *Research Policy* 39(6): 790–798.

Nakagane, Katsuji (2002) *Keizai Hatten to Taisei Iko* [*Economic Development and System Transition*], Nagoya: University of Nagoya Press (in Japanese).

Nakagane, Katsuji (2010) *Taisei Iko no Seijkeizaigaku* [*The Political Economy of Transition: Farewell to Socialism*], Nagoya: University of Nagoya Press (in Japanese).

Nakagawa, Ryoji (2007) *Chugoku no IT Sangyo: Keizai Seicho Hoshiki Tenkan no Naka deno Yakuwari* [*China's IT

Industry: A Role in the Transformation of Economic Growth Way], Kyoto: Minerva Shobo (in Japanese).

National Bureau of Statistics of China (2013) *Zhongguo Tongji Nianjian 2013* [*China Statistical Yearbook 2013*], Beijing: China Statistics Press (in Chinese).

National Bureau of Statistics of China (various years) *Zhongguo Tongji Nianjian* [*China Statistical Yearbook*], Beijing: China Statistics Press (in Chinese).

Naughton, Barry (1995) *Growing out of the Plan: Chinese Economic Reform, 1978–1993*, Cambridge: Cambridge University Press.

Naughton, Barry (2007) *The Chinese Economy: Transitions and Growth*, Cambridge: The MIT Press.

Ning, Lutao (2009) *China's Rise in the World ICT Industry*, London and New York: Routledge.

Numagami, Tsuyoshi, Shigeru Asaba, Junjiro Shintaku and Hisanaga Amikura (1992) "Taiwa to shiteno Kyoso: Dentaku Sangyo ni okeru Kyoso Kodo no Saikaishaku [Competition as an Interaction: Reinterpretation of Competitive Behavior in Electronic Calculator Industry]," *Organizational Science* 26(2): 64–79 (in Japanese).

Ohara, Moriki (1998) "Chugoku Kaden Sangyo no Yuisei: Eakon Sangyo no Sangyo Soshiki to Haiaru Gurupu no Jirei Kara [Advantages of China's Home Appliance Industry: Industrial Organization in the Air Conditioner Industry and a Case Study of the Haier Group]," *Ajiken World Trend* 36: 38–44 (in Japanese).

Ohara, Moriki (2000) "Chugoku Kaden Meka no Kyoso Yui [Competitive Advantages of Chinese Home Appliance Manufacturers]," *Nicchu Keikyo Journal* 2: 6–16 (in Japanese).

Ohara, Moriki (2006) *Interfirm Relations Under Late Industrialization in China: The Supplier System in the Motorcycle Industry*, Chiba: Institute of Developing Economies.

Ohashi, Hideo (2003) *Keizai no Kokusaika* [*Internationalization of Economy*], Nagoya: University of Nagoya Press (in Japanese).

Olley, G. Steven and Ariel Pakes (1996) "The Dynamics of Productivity in the Telecommunications Equipment Industry," *Econometrica* 64(6): 1263–1297.

Ouyang, Shenghai (2006) *Shouji Fenxiao yu Jingying zhi Dao* [*The Way of Mobile-Phone Handset Distribution and Management*], Haikou: Hainan Chubanshe (in Chinese).

Parente, Stephen L. and Edward C. Prescott (2000) *Barriers to Riches*, Cambridge: The MIT Press.

Pecht, Michael, Chung-Shing Lee, Wang Yong Wen, Zong Xiang Fu and Jiang Jun Lu (1999) *The Chinese Electronics Industry*, Boca Raton: CRC Press.

Pei, Changhong, Dexu He and Linbo Jing, eds. (2008) *Zhongguo Duiwai Kaifang yu Liutong Tizhi Gaige 30 nian Yanjiu* [*A Study on 30 years of China's Open-Door Policy and Distribution System Reform*], Beijing: Economy & Management Publishing House (in Chinese).

Popkin, James M. and Partha Iyengar (2007) *IT and the East: How China and India are Altering the Future of Technology and Innovation*, Boston: Harvard Business School Press.

Qian, Qichen (2003) *Waijiao Shiji* [*10 Diplomatic Episodes*], Beijing: Shijie Zhishi Chubanshe (in Chinese)

Qiu, Bin (2009) *FDI Jishu Yichu Qudao yu Zhongguo Zhizaoye Quanyaosu Shengchanlv Zhengzhang Yanjiu* [*A Study on FDI Technology Spillover Channels and TFP Growth of China's Manufacturing Sector*], Nanjing: Southeast University Press (in Chinese).

Rivera-Batiz, Luis A. and Maria-A Oliva (2003) *International Trade: Theory, Strategies, and Evidence*, Oxford: Oxford University Press.

Rivera-Batiz, Luis A. and Paul M. Romer (1991) "Economic Integration and Endogenous Growth," *Quarterly Journal of Economics* 106(2): 531–555.

Romer, Paul M. (1986) "Increasing Returns and Long-Run Growth," *Journal of Political Economy* 94(5): 1002–1037.

Romer, Paul M. (1990) "Endogenous Technological Change," *Journal of Political Economy* 98(5): S71–S102.

Sato, Yukihito (2007) *Taiwan Haiteku Sangyo no Seisei to Hatten* [*The Generation and Development of Taiwan's High-Tech Industry*], Tokyo: Iwanami Shoten (in Japanese)

Sauré, Philip (2007) "Revisiting the Infant Industry Argument," *Journal of Development Economics* 84(1): 104–117.

Sekine, Takashi (2014) *Nihon, Chugoku, Kankoku ni okeru Kadenhin Ryutsu no Hikaku Bunseki* [*Comparative Analysis among Home Appliance Distributions in Japan, China, and Korea*], Tokyo: Dobunkan Shuppan (in Japanese).

Shanghai Caijing Daxue Ketizu (2006) 2006 *Zhongguo Chanye Fazhan Baogao: Zhizaoye de Shichang Jiegou, Xingwei yu Jixiao [2006 Report on China's Industrial Development: Market Structure, Conduct, and Performance in the Manufacturing Sector]*, Shanghai: Shanghai University of Finance & Economics Press (in Chinese).

Shiu, Jingming and Kenichi Imai (2007) "A Divergent Path of Industrial Upgrading: Emergence and Evolution of the Mobile Handset Industry in China," *IDE Discussion Paper* No. 125.

Shiu, Jingming and Kenichi Imai (2010) "Keitai Denwa Sangyo ni okeru Suichoku Bugyo no Suishinsha: IC Meka to Dezain Hausu [Promulgators of Vertical Specialization in the Mobile-Phone Handset Industry: IC Venders and Design Houses]," in Tomoo Marukawa and Masanori Yasumoto, eds., *Keitai Denwa Sangyo no Shinka Purosesu: Nihon ha Naze Koritsu Shita noka? [Evolution Process in the Mobile Industry: Why Japan has been Isolated?]*, Tokyo: Yuhikaku: 197–225 (in Japanese).

Shiu, Jingming, Kenichi Imai and Hirofumi Tatsumoto (2008) "Gijutsu Purattofomu to Seihin Purattofomu no Shijoka: Chugoku Keitai Denwa Sangyo no Kesu [Technology Platform and Product Platform: A Case of China's Mobile Industry]," *MMRC Discussion Paper* No. 226 (in Japanese).

Solow, Robert (1956) "A Contribution to the Theory of Economic Growth," *Quarterly Journal of Economics* 70(1): 65–94.

Suehiro, Akira (2000) *Kyatti-appu gata Kogyoka Ron: Azia Keizai no Kiseki to Tenbo [Catch-Up Industrialization: The Trajectory and Prospects of East Asian Economies]*, Nagoya: University of Nagoya Press (in Japanese).

Takahashi, Goro, ed. (2008) *Kaigai Shinshutsu suru Chugoku Keizai [The Chinese Economy Expanding Overseas]*, Tokyo: Nippon Hyoron Sha (in Japanese).

Tang, Jin (2009) *Higasi Azia ni okeru Nidankai Kyatti-appu Kogyoka: Chugoku Denshi Sangyo no Hatten [Two Stages of Catch-Up Industrialization in East Asia: Growth of Electronics Industry in China]*, Tokyo: Senshu University Press (in Japanese).

Thun, Eric (2006) *Changing Lanes in China: Industrial Development in a Transitional Economy*, Cambridge: Cambridge University Press.

Todo, Yasuyuki (2008) *Gijutsu Denpa to Keizai Seicho: Gurobaruka Jidai no Tojokoku Keizai Bunseki [Technology Diffusion and*

Economic Growth: An Economic Analysis of Developing Countries in the Era of Globalization], Tokyo: Keiso Shobo (in Japanese).

Tse, Edward (2010) *The China Strategy: Harnessing the Power of the World's Fastest-Growing Economy*, New York: Basic Books.

Tuan, Chyau, Linda F. Y. Ng and Bo Zhao (2009) "China's Post-Economic Reform Growth: The Role of FDI and Productivity Progress," *Journal of Asian Economics* 20(3): 280–293.

UNCTAD (United Nations Conference on Trade and Development) (2001) *World Investment Report 2001: Promoting Linkages*, New York and Geneva: United Nations.

UNCTAD (United Nations Conference on Trade and Development) (2012) *World Investment Report 2012: Towards a New Generation of Investment Policies*, New York and Geneva: United Nations.

Vernon, Raymond (1966) "International Investment and International Trade in the Product Cycle," *Quarterly Journal of Economics* 80(2): 190–207.

Wang, Zhile and Jihua Ding, eds. (2012) *2012 Zouxiang Shijie de Zhongguo Kuaguo Gongsi* [*2012 Chinese Transnational Corporations*], Beijing: China Economic Publishing House (in Chinese).

Watanabe, Mariko (2010) "Teishitsu na Seido no moto deno Kigyo Senryaku: Daikin Kaishu Risuku eno Chugoku Kigyo no Hanno ni tsuiteno Keiyaku Riron Bunseki [Good Strategy or Institution?: A Contract Theory Analysis on Response to Default Risk on Trade Credit in Transition China]," *Ajia Keizai* 51(1): 2–30 (in Japanese).

Watanabe, Mariko, ed. (2013) *Chugoku no Sangyo wa Donoyoni Hatten Shitekitaka?* [*How Chinese Industries Have Developed?*], Tokyo: Keiso Shobo (in Japanese).

Watanabe, Mariko and Koichiro Kimura (2012) "Chugoku no Sangyo wa Donoyoni Hatten Shitekitaka? Mondai no Haikei [How Chinese Industries have Developed? Background of the Issue]," in Mariko Watanabe, ed., *Chugoku no Sangyo wa Donoyoni Hatten Shitekitaka?* [*How Chinese Industries have Developed?*], Interim Report, Chiba: Institute of Developing Economies (in Japanese).

Watanabe, Toshio (1979) *Azia Chusinkoku no Chosen* [*Challenges of Asia's Newly Industrializing Countries*], Tokyo: Nikkei (in Japanese).

Wedeman, Andrew H. (2003) *From Mao to Market: Rent Seeking, Local Protectionism, and Marketization in China*, Cambridge: Cambridge University Press.

Williamson, Oliver E. (1985) *The Economic Institutions of Capitalism: Firms, Markets, Relational Contracting*, New York: The Free Press.

Williamson, Peter J. and Ming Zeng (2009) "Chinese Multinationals: Emerging through New Global Gateways," in Ravi Ramamurti and Jitendra V. Singh, ed., *Emerging Multinationals in Emerging Markets*, Cambridge: Cambridge University Press.

Wu, Danhong (2005) *Nanfei Jingji yu Shichang* [*South African Economy and Market*], Beijing: China Commerce and Trade Press (in Chinese).

Wu, Jinglian (2010) *Dangdai Zhongguo Jingji Gaige Jiaocheng* [*A Course on Modern China's Economic Reform*], Shanghai: Shanghai Far East Publishers (in Chinese).

Xie, Xiaoxia (2000) "Kuaisi Zhengzhang de Dianzi Xinxi Chanye [Electronics and Information Industry in High Gear]," in Institute of Industrial Economics, Chinese Academy of Social Sciences, ed. *Zhongguo Gongye Fazhan Baogao (2000): Zhongguo de Xin Shiji Zhanlüe: Cong Gongye Daguo Zou Xiang Gongye Qiangguo* [*China's Industrial Development Report: China's Strategy for the New Century: From a Major Industrial Country to a Powerful Industrial Country*], Beijing: Economy and Management Publishing House (in Chinese).

Xu, Dengfeng (2010) *Zhongguo Qiye Duiwai Zhijie Touzi Jinru Moshi Yanjiu* [*Study on the Entry Mode of Chinese Enterprises' Outward Foreign Direct Investment*], Beijing: Economy and Management Publishing House (in Chinese).

Yang, Lihua, ed. (2010) *Nanfei* [*South Africa*], Beijing: China Social Sciences Press (in Chinese).

Yasumoto, Masanori (2010) "Kaigai no Keitai Denwa Sangyo no Tenki: Gurobaru na Gijutsu no Fukyu to Shijo no Bunka [The Turning Point of Overseas Mobile Industries: Globalized Diffusion of Technology and Differentiation of the Market]," in Tomoo Marukawa and Masanori Yasumoto, eds., *Keitai Denwa Sangyo no Shinka Purosesu: Nihon ha Naze Koritsu Shita noka?* [*Evolution Process in the Mobile Industry: Why Japan has been Isolated?*], Tokyo: Yuhikaku: 129–172 (in Japanese).

Yin, Xingmin (2003) *Zhongguo Gongye yu Jishu Fazhan* [*China's Industry and Technology Development*], Shanghai: Shanghai Joint Publishing Press and Shanghai People's Publishing House (in Chinese).

Young, Alwyn (1991) "Learning by Doing and the Dynamic Effects of International Trade." *Quarterly Journal of Economics* 106(2): 369–405.

Yuan, Zhijia (2009) "Kokuyu Kigyo no Seisei, Shinka, Henkaku Katei: Kasho Denshi no Jirei kara [Generation, Evolution and Transformation Processes of State-Owned Enterprises: From a Case of Huajing Electronics]," in *Gendai Chugoku Kigyo Henkaku no Ninaite: Tayoka suru Kigyo Seido to Sono Shoten* [*Leaders of Contemporary Chinese Firms Transformation: Diversifying Firm Institution and Its Focus*], Tokyo: Hihyosha: 72–89 (in Japanese).

Yuan, Zhijia (2013) "ASEAN ni okeru Chugoku Takokuseki Kigyo no Sangyo Kyosoryoku no Hikaku Bunseki: Jidosha to Denshi Sangyo o Chushin ni [Comparative Analysis of Industrial Competitiveness of Chinese Multinational Enterprises in ASEAN: With Focuses on Automotive and Electronics Industries]," in Shinichi Kawai, ed., *Chugoku Takokuseki Kigyo no Kaigai Keiei* [*Overseas Operations of Chinese Multinational Enterprises*], Tokyo: Nippon Hyoron Sha: 85–105 (in Japanese).

Zhang, Kevin Honglin (2009) "Rise of Chinese Multinational Firms," *The Chinese Economy* 42(6): 81–96.

Zhao, Ying (2000) *Zhongguo Chanye Zhengce Shizheng Fenxi* [*Empirical Analysis of Chinese Industrial Policy*], Beijing: Social Sciences Academic Press (in Chinese).

Zheng, Ruihong (2012) *Yingxiao Qudao Guanli* [*Marketing Channels Management*], Beijing: China Machine Press (in Chinese).

Zhongguo Dianzi Gongye 50 nian Bianweihui (1999) *Zhongguo Dianzi Gongye 50 nian* [*50 Years of China's Electronics Industry*], Beijing: Publishing of Electronics Industry (in Chinese).

Zhongguo Xinxi Chanye Nianjian Bianweihui (2010) *Zhongguo Xinxi Chanye Nianjian (2010)* [*Yearbook of China Information Industry (2010)*], Beijing: Publishing House of Electronics Industry (in Chinese).

Index

advantage of backwardness, 16, 17, 19, 20, 135
African National Congress, 118
Alcatel, 97
AMAP, 125
Amoi Electronics (Amoi), 84, 97, 100
automotive industry, 27, 56
away disadvantage, 6

Bastable's criterion, 22
BenQ, 89
Bertrand competition, 108, 109, 146
boundaries of the firm, 14, 34–5
brand recognition, 95, 116, 117, 118, 124, 125, 129

China Mobile, 143
China Unicom, 143
closed-door policy, 1
Cobb-Douglas production function, 62, 104
Code Division Multiple Access (CDMA), 78, 143
Cold War, 25
Communist Party of China, 118
comparative advantage, 21, 26–7

comparative statics, 14
complete-knockdown (CKD), 124, 148
concentration ratio of the top ten firms in sales ($CR10$), 71, 73
consumer electronics products, 4, 37, 45, 119
core layer, 87, 88, 98
core technologies, 5
Cultural Revolution, 45
Czech Republic, 21

defense industry, 45–6, 47
Defy, 125, 126, 127
Dell, 90
dependency theory, 21
developed countries, 1–2
developing countries, 1–2
development, 5
digitalization, 6, 28
disintegration, 8, 32, 107, 139
diversification mechanism, 7–8
dynamic economies of scale, 22

East Asia, 17
economic reform and open-door policy, 3, 24
electrical industry, 4
electronics industry, 4
endogeneity problem, 20, 63

endogenous growth theory, 14–15
Ericsson, 80, 144
  see also Sony Ericsson
European Free Trade Area (EFTA), 122
European Union (EU), 88, 97, 122, 123
experience effect, 22, 65
export-oriented industrialization (EOI), 3
externalities, 22, 23
external technology, 7

Five-Year Plan
  First, 45
  Seventh, 142
foreign firms, 1–2
foreign loans, 43

Game Stores (Game), 125, 126
generalized least squares (GLS), 68
globalization, 2
global knowledge, 7
Global System for Mobile Communications (GSM), 78, 87, 91, 143
Going Global (*Zouchuqu*), 11, 116
Gome, 94
Gree Electric Appliances (Gree), 27, 46
gross industrial output value, 38, 40
growth mechanism, 37

Haier Group (Haier), 27, 30, 46, 77, 122, 137, 147
hardware, 87, 88, 97, 98

Harmonized Commodity Description and Coding System (HS), 38, 52, 53, 142
headquarters service-intensive industry, 108, 111, 112
hei shouji (black hansets), 94
  see also illegal handsets
heterogenization, 133
Hewlett-Packard, 90
hierarchical system of technologies, 33
Hisense Group (Hisense), 27, 120, 121, 122, 125, 126, 127, 128
hold-up problem, 34–5, 105
home advantage, 6
home appliances, 4
homogenization, 11, 114, 132, 133
Hong Kong, 3, 18, 145
Hon Hai Precision Industry (Hon Hai), 89, 90
Huawei Technologies (Huawei), 27, 78, 82, 84, 121, 122, 137
human capital, 15, 34, 35, 103, 105, 106, 108, 140

illegal handsets, 83, 94, 144
  see also hei shouj; *shanzhai ji*
imitation, 5, 9, 17, 18, 19, 20, 57
implicit knowledge and know-how, 19
import-substitution industrialization (ISI), 3, 48
incomplete contracts, 34–5, 102

independent design houses (IDHs), 79, 89, 96, 97, 99
India, 20, 47–8
indigenous firms, 1–2
industrial agglomerations, 117, 141
Industrial Classification for National Economic Activities, 37
industrial policy, 81, 90, 94
Industrial Revolution, 22
infant industry, 21–3
information and communication technology (ICT), 38, 90, 121, 122
innovation, 17, 141
integration, 8, 32, 35, 107, 111
vertical, 139
internal knowledge, 7
international division of labor, 27, 42, 49
interventions, 16

Japan, 17, 46, 55, 80, 90, 117, 124, 144

Kemp's criterion, 22–3
key components, 5, 6, 10, 28, 29, 42, 49, 51, 77, 79, 81, 85, 87, 104, 106, 110, 112, 125, 131
KIC, 127
Konka Group (Konka), 27, 83, 84, 88, 91, 121, 122, 124, 126

labor-intensive industries, 26
large conglomerates (*chaebol*), 17
latecomers, 6, 22, 33, 65, 133

learning-by-doing (LBD), 15, 19, 23, 33, 35, 65, 103
learning-by-exporting, 18
learning effect, 22, 65
Lenovo, 27, 77, 83, 84, 86, 96–7, 98–9, 121, 122, 137
LG Electronics (LG), 89, 124, 127
liberalization, xiv, 4, 25, 27, 31, 45, 47, 48
licensing system, 47, 89, 144
Liu, Chuanzhi, 77–8
local knowledge, 7

make-or-buy, 4–5
Malaysia, 3
Mandela, Nelson, 118
manufacturing, 5
marginal costs, 107, 109
market failure, 22–3
marketing-oriented strategy, 28, 91, 92, 95, 96, 136
market-stealing effect, 3, 21, 33, 68, 142
see also negative effect
market structures, 47
Matsushita Electric Industrial, 46, 80
see also Panasonic
MediaTek (MTK), 97–8, 99, 100
mercantilism, 21–2
Mexico, 21
middle layer, 87, 88, 98
Midea Group (Midea), 27
Mill's criterion, 22
modularization, 6, 28, 88, 94, 112, 141
Motorola, 80, 84, 90

multinational enterprises (MNEs), 18, 29, 116, 117, 118, 128, 129, 147, 148

nationalization project, 81, 89, 103
NEC, 80
negative effect, 3
  see also market-stealing effect
neoclassical growth theory, 15
newly industrializing economies (NIEs), 17, 18
Ningbo Bird (Bird), 82–3, 84, 89, 91, 92, 93, 97, 98
Nokia, 80, 84
North American Industry Classification System (NAICS), 58, 60

open-economy models, 15
openness, 3, 15, 20, 24, 25, 26, 139, 140
organizational form, 4–5, 7–8, 11, 45, 131, 132
original design manufacturers (ODMs), 89, 96
original equipment manufacturers (OEMs), 27, 89, 96, 125, 127, 128, 129
outside firms, 6
outside option, 105
overseas expansion, 1, 115, 137

Panasonic, 46, 144
  see also Matsushita Electric Industrial
Pantech, 89, 90

patents, 19, 25
Pearl River Delta, 26
Philips, 80, 84
Pick n Pay Stores (Pick n Pay), 125, 126
planned economy era, 4, 26, 30, 45
platforms, 29, 88, 97–100
positive effect, 3
  see also technology spillover effect
productivity gap, 15, 66
  see also technology gap
product lifecycle (PLC), 18–19, 136
protectionist policies, 17, 31, 48

Quanta Computer (Quanta), 89, 90

reference designs, 97–8
relationship-specific investment, 34
research and development (R&D), 5, 15, 16, 19, 26, 49, 56, 81, 97, 115–16, 118, 121, 124, 132, 137

Sagem, 97
sales, 5
sales and after-the-sale service networks, 6, 28, 77
  see also sales networks
sales networks, 10, 28, 29, 79, 91, 93, 95, 142
  see also sales and after-the-sale service networks
Samsung Electronics (Samsung), 84, 124, 127

INDEX

second generation of mobile phone standard (2G), 78
semi-knockdown (SKD), 123, 124
Sewon Telecom, 89
shanzhai ji (bandit handsets), 94
   see also illegal handsets
Sichuan Changhong Electronic (Changhong), 27, 46, 122
Siemens, 80, 84
Singapore, 3, 18
social ability, 17
software, 29, 50, 79, 87, 88, 97, 98
Sony, 90, 124, 127, 144, 148
Sony Ericsson, 84, 144
   see also Ericsson
South Africa (SA), 11, 117
Southern African Development Community (SADC), 122, 123
South Korea (Korea), 3, 17, 18, 48, 55, 85, 89, 117, 120, 124, 140
Soviet Union, 25, 45
specialization in exports, 51
specialization in imports, 51
spillover externalities, 22
state-owned enterprises (SOEs), 46, 119
store brands (SBs), 125, 127
structuralism, 21
Suning, 94
sunk investment, 34
surface layer, 87, 88, 98
SVA Group (SVA), 119, 120, 121, 122, 124, 125, 126, 127

Taiwan, 3, 18, 85, 89, 145
tariff-jumping investments, 119, 122
TCL, 27, 83, 84, 89, 91, 93, 97, 98, 121, 122, 126, 141, 147
technological capabilities, 16, 78, 81, 82, 85, 88, 89, 90, 97, 99, 100, 103
technological change, 15, 26
technologically advanced goods, 113
   see also technologically advanced goods and services
technologically advanced goods and services, 5–6, 13
   see also technologically advanced goods
technological obsolescence, 18–19
technology, 2
technology absorption, 19, 26
technology acquisition, xiv, 2, 5, 27, 95, 96, 133, 134, 135
technology diffusion, 2–3
   incomplete, 13, 32
technology gap, 5
   see also productivity gap
technology imports, 2
technology introductions, 2
technology service-intensive industry, 105, 108, 110, 111, 112, 114
technology spillover effect, 3
   see also positive effect
technology transfers, 2
Tedelex, 121, 125, 127

telecommunication carriers, 79, 94, 121, 143, 145
Telefunken, 127
Texas Instruments, 88
Thailand, 3, 17
third generation of mobile phone standard (3G), 78, 143
Time Division-Synchronous Code Division Multiple Access (TD-SCDMA), 78, 143
total factor productivity (TFP), 9, 26, 32, 55
township-and-village enterprises (TVEs), 46
trade deficits, 42
trade policies, 20, 23, 124, 140
trade protection, 140
trade specialization coefficient, 49
trade surpluses, 42

unforeseen contingencies, 34, 142
United Kingdom (UK), 22, 144
United States (US), 16, 17, 22, 25, 80, 88, 90, 97, 140

value added, 38, 40, 46, 50, 56, 60, 61, 63, 67, 68, 70, 74, 80
value chains, 4–5, 32, 49, 139
Venezuela, 21
vertical specialization, 6, 9, 28, 29, 31, 37, 48–9, 51, 94, 112, 117, 133, 139

world's factory, 4, 38
World Trade Organization (WTO), 25, 31, 144

Xiamen Chinese Overseas Electronic (XOCECO), 27, 120–1, 122, 124, 126, 127

Yangtze River Delta, 26

zero boundaries, 111, 114, 134
    see also zero organizations
zero organizations, 11
    see also zero boundaries
ZTE, 27, 82, 84, 122, 137

GPSR Compliance
The European Union's (EU) General Product Safety Regulation (GPSR) is a set of rules that requires consumer products to be safe and our obligations to ensure this.

If you have any concerns about our products, you can contact us on

ProductSafety@springernature.com

In case Publisher is established outside the EU, the EU authorized representative is:

Springer Nature Customer Service Center GmbH
Europaplatz 3
69115 Heidelberg, Germany

www.ingramcontent.com/pod-product-compliance
Lightning Source LLC
LaVergne TN
LVHW011007250326
834688LV00004B/113